Roger Green is Director of the Centre for Community Research at the University of Hertfordshire in Hatfield, and is a regular contributor to academic publications. He lives in East London. *Destination Nowhere* is his first book.

DESTINATION NOWHERE
*A South Mimms Motorway
Service Station Diary*

DESTINATION NOWHERE

A South Mimms Motorway Service Station Diary

Roger Green

ATHENA PRESS
LONDON

DESTINATION NOWHERE
A South Mimms Motorway Service Station Diary
Copyright © Roger Green 2004

All Rights Reserved

No part of this book may be reproduced in any form
by photocopying or by any electronic or mechanical means,
including information storage and retrieval systems,
without permission in writing from both the copyright
owner and the publisher of this book.

ISBN 1 84401 352 9

First Published 2004 by
ATHENA PRESS
Queen's House, 2 Holly Road
Twickenham TW1 4EG
United Kingdom

Printed for Athena Press

Acknowledgements

A big thanks to the staff of Welcome Break at South Mimms Service Station who agreed to talk to me; particularly Matthew Chilcott and Stewart Ashby of Welcome Break for making it happen; David of Coffee Primo fame for his excellent café lattés; motorway travellers who gave me their time in between nipping to the toilets and cups of tea; the residents of South Mimms, past and present, and others connected with the service station and the site who agreed to give me their time.

Without Sharon Curtis's help in typing up my observation diaries and notes, and interview tapes, this book would have not been completed. Thanks as well to Sharon Forde for gathering useful information by searching the Internet.

A special mention to Willie at the Hair Bus Company, for the haircuts and conversations; David Ranger, from the University of Hertfordshire, who risked life and limb taking the photographs; Muriel Brittain for the loan of her late husband's material relating to the history of South Mimms village and the surrounding area.

Finally thanks to my mate, Jim, for his text messages, who worryingly at times has become as obsessed by the service station as myself.

Most of the names in the book have been changed.

Contents

1: Introduction	13
2: Becoming a Regular	18
3: 'I think he might be Lord Lucan'	43
4: Shaggers	58
5: I was a Social Worker	83
6: Breakfast at the Red Hen	101
7: 'Hello Tel'	125
8: Trouble with the Saga Louts	153
9: Bermuda Triangle	169
10: 'It's all happening here' – England v. Nigeria	192
11: Preacher Man	210
12: A Few References	225

Motorway Services: An area beside a motorway providing facilities for motorists, usually including a petrol station, cafeteria, and shopping service.

Oxford English Dictionary

1: Introduction

> I realise that it is your job as a reader to try to create a credible consistent pattern out of all this.
>
> *Luke Rhinehart*

Not sure how it all started. I seem to have found myself travelling around the M25 one morning heading west away from an early morning red sun, and decided that I was too young to die after a forty-four tonne articulated lorry the size of Belgium tail-gated me all the way from Junction 25 at the A10 to pass the Potters Bar turn-off at Junction 24. At the blue and white rectangular information sign approaching Junction 23 I made an emergency exit for South Mimms Services and found my way on to a congested tarmac littered with cars and white transit vans.

I had been there a couple of times in the past for a quick overpriced coffee and to use the toilet en route to somewhere on the south coast before a fire in Julie's Pantry restaurant burnt the place down in 1998. Motorway service stations had never been a big thing with me, just places you pass through. This was different. A regular café latté and a pain au chocolat whilst reading the paper was somehow civilised. No roar of the motorway traffic. My driving gloves were off and I slipped into rest and recovery mode. I was in an oasis. I had entered the world of the motorway service station.

I started to read up about the history of the motorway service station in the UK and discovered to my surprise that they had been around since the early 1960s when the Watford Gap and Newport Pagnell service stations opened on the M1. Originally centrally planned and controlled, and situated at minimum intervals of 30 miles, they were seen as self-contained units – alcohol free zones, providing 24-hour catering for private car owners, coaches, and lorry drivers. In addition, they provided toilet and refuelling facilities for 365 days a year. The intention was to provide highly regulated watering holes in the middle of nowhere.

As a nation of motorway road users, we seem to visit them more often than we go to church on a Sunday. Surrounded by utilitarian décor we stop at them to stretch our legs, make a beeline to the toilets, and to spend the total Gross Domestic Product of a developing country on refreshments and the purchase of meaningless tacky items from tourist style shops.

Motorway service areas such as Clacket Lane, Fleet, Heston, Newport Pagnell, Scratchwood, South Mimms, Toddington, and Watford Gap have stuck in the public consciousness and became part of our popular culture. They are contemporary equivalents of the coaching inn – familiar landmarks we use to negotiate motorways as we hurtle through the countryside.

They exist to process motorway users with minimal delay and investment whilst at the same time providing an acceptable rest stop with basic amenities. The car driver, as Nigel Richardson remarks in his foreword to Jon Nicholson's photographic journey up the A1, sticks to them like flypaper. Just like at mainline railway stations and airport terminals, we linger for a while, peruse their delights, before continuing onward to our destinations.

Officially opened in 1987 by Mrs Thatcher, South Mimms reminded me of Alain de Botton's description of the service station in his book *The Art of Travel* as 'like a lighthouse at the edge of the ocean, it seemed not to belong to the city, nor to the country either, but rather to some third, traveller's realm'.

Bordered in the Hertfordshire countryside to the west by the wonderfully named Trotters Bottom and a travellers' caravan site hidden behind modern earthworks and to the north-west by the eighteenth-century Clare Hall, an ex-private smallpox hospital, and a sanatorium for tuberculosis patients, now a Cancer Research Centre with high security fences and 'keep out' notices. To the north the picturesque village of South Mimms, first mentioned in the Doomsday Book of 1086, and with its own Women's Institute, looks down on it. To the south-east the sprawling town of Potters Bar is just visible across cornfields with burnt-out cars dumped in them by joyriders. There are no footpaths to the service station. You drive in and drive out. The service station is somehow not part of the real world. It fell to earth like David Bowie.

Yet staff, along with their customers, somehow provide service stations with a sense of community. Visiting local residents add to this feeling. This public space designed to process motorway users in twenty minutes through the mechanically efficient automated sliding entry and exit doors, becomes personalised. Service stations, places deemed by government regulations as stopovers but not destinations in their own right, develop a life of their own. 'On the way to becoming a place', wrote David Lawrence, in *Always a Welcome. The glove compartment history of the Motorway Service Station*, an account of his travels up and down Britain's motorways in the 1990s.

From the outside South Mimms has a ribbed aluminium roof kept clean by the service area's own microclimate. Everywhere large white swans seemingly glued to the external walls fly across your vision whilst oversize posters advertise the goodies that await you inside. Visit at night, and blue neon strip lighting hugs the edge of the silver coloured aluminium sheet roof like an advertising sign out of the set of the sci-fi film *Blade Runner*.

You are led through the entrance under a raised canopy roof in the form of a wave. Your senses become overloaded as you enter. Inside there is shiny glass everywhere. Light floods the foyer and the concourse. It's a cross between Bluewater shopping centre and the departure lounge of Gatwick Airport. Or is it an aircraft hanger? Or a cathedral promoted by some young trendy Church of England vicar? The motorist is being asked to momentarily forget the stresses of the road and indulge in a fantasy world.

Past the Game Zone, the Coffee Primo, and the Red Hen restaurant on the left with its All Day Brunch, and Van Gogh's tiny Workroom on the right, you are faced with The Granary. Signage indicates that you can purchase instant snacks such as soup or sandwiches, or a pizza, or, if you really feel up to it, a traditional breakfast for £6.49. Further round, the two red and

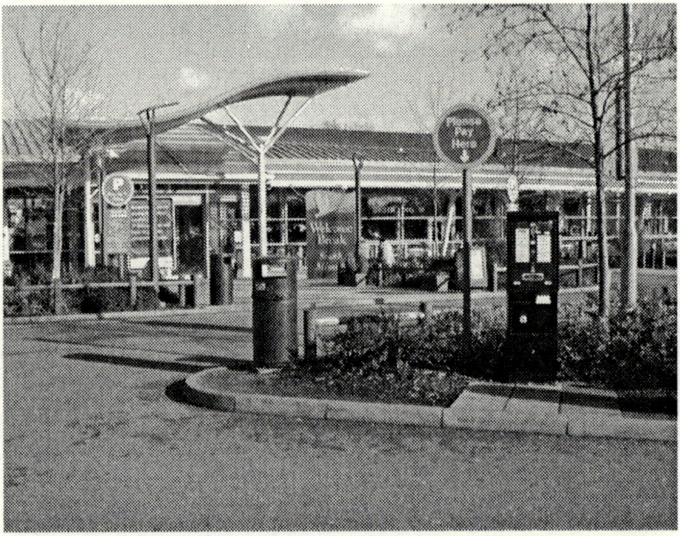

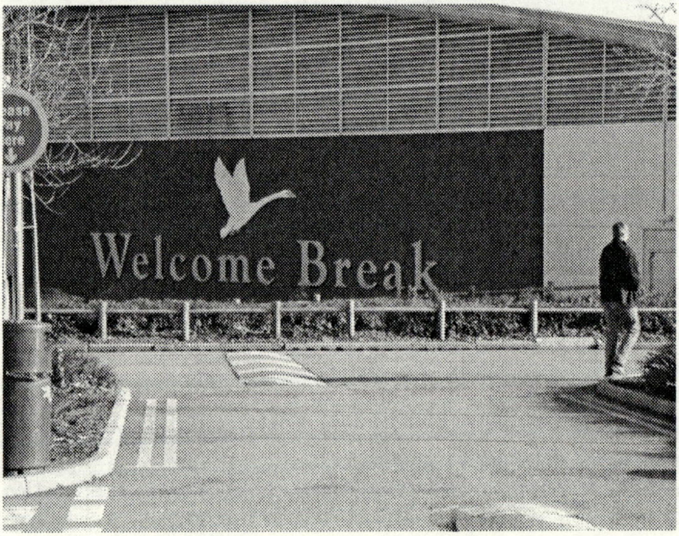
Welcome Break

white decorated fast food outlets, Burger King and Kentucky Fried Chicken, beckon you before you hit the segregated smoking area.

Going to the toilet is no escape from this visual free-for-all. Blown-up photos of plates of bangers and mash, meat pie, and fish and chips are there to remind you that you are hungry and you do need to eat.

Finding somewhere to sit with your overpriced meal is not a problem. There are over 250 seats in the no-smoking eating area and some 100 plus in the carpetless smokers' corner. With the furniture being a mixture of IKEA and Habitat tables and chairs, high stools, and chrome and wicker chairs, it's got a comfy coffee shop feel to it.

As you eat, your eyes meet the sign above the shop which informs you that 'Welcome Break South Mimms' is 'The World's finest Motorway Service Area'.

I began to see my fellow service station users as individuals. Maybe breaking motorway journeys in architecturally designed film sets with their interiors of garish colours, harsh lighting, and hidden CCTV surveillance cameras allows people to be themselves. A modernity where we share space with strangers in a public place free of the constraints of our normal surroundings with their familiarity.

Call it voyeuristic curiosity but I could feel myself becoming entranced by the seeming chaos and order going on all around me. Like a bar out of a *Star Wars* movie the place was full of strange and exotic people. It gave me the promise that if I hung around enough it could be unravelled over time so that I might begin to see what was happening. It was late December 2000; I was hooked. Fuelled by coffee, I became a 'regular'.

2: Becoming a Regular

> This is a great moment, when you see, however distant, the goal of your wandering. The thing which has been living in your imagination suddenly becomes part of the tangible world.
>
> *Freya Stark*

Not sure what to do – pass through the metal detector – travellers' camp – meet the table clearers – Marks and Spencer's fleeces – toilet noises – the builders are in – Happy New Year – Kathmandu revisited – 'Posh' fans meet the Birmingham County Netball Team and the England Rugby XV – Judith the GP – my *Guardian* is mopped – nearly get arrested – Karl Marx is in – meet Sadaam Hussein in the car park – flash geezer – I'm invisible – the price of chocolate croissants is going up – the fat controller makes an appearance.

18 December

It's 7.45 a.m. I'm new. I have trouble finding the gents', probably because I have to pass through a metal detector en route. Decide I need a coffee. Two staff, Michelle and Julie, are waiting for me at The Granary counter. I join a disorganised queue and nervously ask for a pot of coffee, uncertain as to the correct procedure.

Michelle slowly moves the heated sausages before filling my pot. The weekend news is being discussed.

'Who's on this morning?' asks Julie.

'That's £1.59 please, love.'

I smile through my teeth. It's a bargain.

Her chubby fingers place the change into my hands.

People are sitting at tables by themselves. We all face towards the entrance; I am half expecting an air stewardess to appear and give the routine emergency life-jacket/evacuation demonstration you get on package holiday flights to Ibiza.

The man sitting beside me noisily chomps his toast. He's dressed in a Mohammed Al Fayed double-breasted grey suit.

Across the way a woman in a red and black fleece leaves her *Daily Mail* on the table. Like a vulture, a businessman plucks the newspaper off the table and carries it to his, two tables up.

One of the army of 'table clearers' picks the bones from her table. It is now clear, with the no-smoking sign in place ready for the next customer.

Mobile phones are ringing two to a dozen. What are people looking at when speaking into their phones? Plenty of facial expressions and hand gestures. I always feel I want to join in the conversations.

The house music pipes the Bee Gees classic song 'How deep is your love' from the *Saturday Night Fever* album, across the tables. This is good!

Feel the twang of an American mid-west accent from a young boy behind me. The adult with him listens intently; both carry mobile phones which are waved about as they shout at each other. Is all human life here?

19 December

Back in the following day. It's a foggy damp morning. I am guided to the entrance by the blue neon light glowing through the gloom. Marks and Sparks fleeces much in evidence this morning. Groups of them cluster over teapots. Must be a coachload en route to an over-fifties event. Follow a group of pensioners with Mancunian accents into the shop. The women make cooing noises around the £14.99 fluffy lions and dogs, and poke the Swiss style bird-houses with withered fingers. Men smoking my late father's favourite Old Holborn tobacco he used in his roll-ups stand around talking, enjoying the moment away from their wives.

Morning ablutions, flushing water, slamming doors, the roar of hand dryers, toilet paper being violently pulled from holders. Silence in the cubicle to my left! Shadows on the floor as men come and go.

'I've got piles.'

'The coach leaves in half an hour, shall I tell the driver to wait?'

A long, strained fart punctuates the air. I decide to move on.

A funny-looking young lad, in his early thirties, sticky-out ears, is in the shop with his mum buying one of the tabloid television news guides.

'You know what's happening two weeks before it happens in the soaps,' she informs a middle-aged black man who is with them but somehow distant.

The orderly queue to pay at the counter builds. Sue is having a bad morning; someone's credit card is not producing the necessary whirl of the acceptance notification. Chris opens up the till alongside her. She looks confused as my *Guardian* newspaper is in two parts.

'It's all one newspaper,' I helpfully inform her.

Three African males emerge from the darkness of the car park outside. In line they stroll through and disappear via the 'Staff Only' door. One has a pink scarf around his neck. I wait but they do not return.

A large-bellied, ponytailed, fifty-something-year-old hobbles through the entrance's sliding doors, blinks and returns outside for a smoke. His black leathery trousers make a swishing noise in tune with the doors as they open and shut.

I am happy drinking my coffee. Christmas songs fill the air around me. I can feel the Christmas spirit. People around me seem oblivious. Does anyone hear them over the mobile phone talk and the rustling of newspapers?

Two young white people, left over from the Glastonbury Festival enter. Serious dreadlocks, dressed in statutory green combat trousers coated with mud and vomit, and tee shirts which were once light coloured. They both sport nose and ear rings. They take coffee with an older man, in his late sixties, with a tin of tobacco by his side. A muted conversation; the young woman looks bemused; her companion simply stares at the grey-haired man. They move off in different directions.

The builders are in. Brown steel-toe capped boots, tartan jacket shirts over white tee shirts and blue jeans. Surprisingly, only one with a ponytail. Looks like the Fat Boy breakfast specials all round. Two of them clutch their mobile phones and practise sharp intakes of breath as they await a call for the next estimate.

*

Accompanied by Lord Forte, Chairman of Trusthouse Forte, the Right Honourable Margaret Thatcher officially opened the first service area on the M25 motorway at South Mimms on Saturday, 6th June 1987.

*

20 December

A gorgeous blond woman dressed in black, red lips, tight-fitting trousers is in. She eyes the men coming out of the gents' toilet. She waits patiently for her husband and her young son, a small child aged about eight.

An older man, with greying hair, but dressed in a three-quarter length expensive black leather coat, sits with a much younger woman. Deep in conversation, his hands clasped together as he leans towards her.

Only one fleece around this morning, but a few middle-aged couples drinking coffee together, 'boys' one side of the table, 'girls' the other side.

The hum of mobile phones chattering.

All white customers – is this usual?

1 January, New Year's Day

'Happy New Year.' The car park is busy, everybody out and about. Feels like a Sunday with people window-shopping and children running about excitedly. They stop and gape up and around at the signs and colours. Clothes given as Christmas presents are being shown off.

Young men promenade with their dazzling blond girlfriends. High heels clatter across the wooden floor.

Women lead men to their tables, the latter carrying expensive meals. Burger King are doing maximum business. For some reason sausages and chips with Coca-Cola are being consumed in large quantities.

A man shouts from somewhere to turn off a mobile phone.

John and his team of table clearers tackle the leftovers on the tables. No longer do they wear their Christmas hats.

A tall Nigerian security guard patrols the aisles, phone held behind his back; his blue uniform stands out from the Christmas colours of male customers.

There is a total absence of 'business people' with their laptops, black cases of files, and their urgency.

A new machine in the Game Zone draws young children to it. It's short-lived. Concerned parents are vigilant. Screaming and crying children are pulled away from its random squirts of noise and glittering lights.

Are people travelling from A to B or simply getting some air? Going for a stroll? Is this the only social gathering place in the locality which is open on New Year's Day?

They are now five deep at the Burger King outlet. Trays of food are ceremoniously carried by dads towards awaiting tables. Tribes of north Londoners tuck in.

A middle-class family of two adults and two small children on the table in front of me dine on Perrier water and sandwiches. They sit and eat quietly. The two children's eyes flicker enviously at the 'hamburger' and milkshakes the other children are enjoying.

2 January

The car park is virtually empty. The newly arrived travellers' encampment of caravans and good second-hand cars remains in darkness. Red gas bottles stand to attention alongside the vans. A couple of Land Cruisers view the proceedings.

Inside is quiet.

Richard with the goatee beard takes my 50p for the *Guardian* newspaper.

I decide to give the traditional £6.99 breakfast a miss this morning.

Rows of empty seats stretch out beyond me like some Anglican Sunday morning service. The Christmas tree by the entrance glistens with its white lights and trimmings. People come and go without a second glance. The La Brioche Dorée, 'a taste of France', is as empty as Blackpool beach on a rainy summer's day. A lone smoker puffs away in their designated section by the toilets.

Despite the 'night after' feel to the place, the sliding doors at the entrance cough out incomers and outgoers with mechanical efficiency.

The Burger King lies dormant, its lights have that half dimmed feel of a bereavement parlour. The 'Official Grand Prix' merchandise shop remains shuttered and closed. The Game Zone next to it offers TV screens of games but only passing shadows take any interest.

Outside the car park is becoming alive with white transit vans from Derby and Stevenage. A Japanese businessman takes time negotiating the boot of his Essex Man's red Ford Mondeo.

The first coaches of the day are arriving. A Welcome Break employee in his blue uniform is doing the rounds. He collects customer rubbish in his black bin-liner. A black used condom is retrieved from the branches of one of the bushes.

*

Strengthened guidelines will ensure the motorway service stations continue to serve the genuine needs of motorists and do not become destinations in their own right. (Highway Agencies Public Relations Department, 1998)

*

3 January

I am greeted by a pack of small dogs running wild in the car park. Their home is the travellers' site, now short on vehicles at this time of morning but full of the stench of human faeces reminiscent of the banks of the Bagmati River, in the Nepalese city of Kathmandu.

The businessmen are slowly returning from their New Year break. They cluster like hungry sparrows around some of the tables. The early morning sound of their mobile phones ringing is a sign that normality is returning. Although a middle-aged couple sporting matching off-red fleeces are a reminder that the holiday season is not quite over yet.

I peruse the Game Zone. It has three customers. A man, late twenties, is busy zapping human images with an AK47 lookalike.

'See that, got the little fuckers.' I nod and ask where he's from.

'Nowhere, mate, and ain't going anywhere neither.'

An older man (his father?) dressed in a colonial Marks and Sparks suit looks on. They are joined by a youth, with a white baseball cap and dozy features.

We stand round and watch.

A customer announcement – first one I have heard – informs the owner of a blue BMW car that they should proceed to the car park where a security officer will meet them. I'm left alone.

*

'Delays on the M25 between Potters Bar and Cheshunt on the A10 due to an accident. Traffic is stretching back to South Mimms services. Thanks to Bill the Bricklayer for that.' (Radio 5 Live Traffic News)

*

4 January

Men are sitting alone at tables. One starts to read a book. Looks like he is passing through onwards to a photographic or chess conference. Funny how men sit at tables at least a metre apart from each other, like some kind of territorial game.

I see my first Asian family, mum and dad, late forties, with their son, in his twenties. They cautiously move towards the 'Big Breakfast' counter. Croissants and coffees emerge. The woman, dressed in a sari, has that elegant sophisticated look to her. Her husband dressed immaculately in a dark grey suit and black scarf, nods as his wife and son talk. They sit away from the rest of us.

The smokers' section is relatively full for a change. Quite a smattering of puffing businessmen and women. From where I am sitting they look like they are somewhere else. In another service station perhaps?

The table clearers stand impassively waiting to swoop. Dressed in their 'Fred Perry' tops and dark trousers, they have a mixture of European languages.

Three table clearers together. Now all three survey us. I feel a collective scan over me, checking if I have finished with my coffee. A fourth table clearer joins them, blue J-cloths at the ready. They disperse, whilst one of them disappears behind their small, walled enclosure. Much clashing of knives and forks, trays being emptied, crockery being stacked on a stainless steel sink, food leftovers being shovelled into the bin. Two emerge. Silence from within the enclosure.

Lots of umbrellas in evidence this morning. Vertical rain this time of morning seems to turn people into faceless zombies as they emerge into the lights of the 'cathedral'. Arms, umbrellas and occasionally legs being shaken as they walk in.

Three Marks and Spencer dressed middle-aged men join me at the next table with Big Breakfasts. Their new jumpers still smell of Christmas wrapping paper. Conversation is muted, in between mouthfuls of egg, bacon and fried bread. Small packets of tomato and brown sauce stain the table.

The Asian family leave; the mother dutifully walks behind her husband. The son leads the procession through the tables and beyond.

A *Daily Telegraph* reader peers over the top of his paper, his half-finished bottle of orange juice and empty coffee pot stand on his tray.

'I said to John, if the deal continues to be discussed in this way...'

The volume is turned down as two businessmen whisper that David is prevaricating.

A broad Lancashire accent explains to his friend that it should be another three to four hours before they get home. Both are Pakistani Muslims from Bradford. One is soberly dressed in casual wear and sports a goatee beard with his black hair. His friend sitting facing him wears a white baseball cap, trainers, and multi-coloured jogging bottoms. His black Adidas top hides a white vest. As he looks through his *Sun* newspaper he asks his friend if he wants anything else to eat. An impassive response, he merely gazes at what is happening around him.

You're not from around here? I am thinking.

Communication is in English, slow and with much gesticulating, his friend fumbles through some sentences.

'Tell them what happened.'

'I say date.'

'No.'

'Forget that, they will ask you questions about how you got here. Remember what you said to me.'

Much staring into space.

The second pot of coffee has activated my water works. Need a pee.

I am stopped on my way to the gents'.

'Would you like a charity credit card, sir?'

'No, thanks, I have enough for my own pack of cards!' I smilingly replied, quickly realising that the two bubbly blond salespersons were directing their questions at a suited businessman behind me. I blush and continue to mutter to myself as I escape to the confines of the gents'.

6 January

Football supporters are everywhere. It's Saturday, of course, when normal everyday reality is suspended and 'football reality' takes over. Fans down the A1M from Peterborough – a group of some fifteen men standing around eating cakes: grown men in football shirts, tummies hanging over their trousers, Peterborough accents laced with a heavy Londoner ring. The 'No Smoking' ban obviously does not apply to them. North of Watford men, noisy, playing away from home, laugh and joke. Much pushing and shoving.

More 'posh' supporters emerge from the sunlight. A fat woman, young, wearing her shirt, and a wiry man. They are travelling to London to see their team play at Chelsea in the third round of the FA Cup. Burger King is doing a brisk trade in refuelling them. Milkshakes, burgers, and large French fries are consumed. Belches and farts join in with their chatter.

The Birmingham County Netball Team arrive in their red tops, blue trousers and trainers. Is Birmingham now a county? A multi-racial group. 'Brummie' accents punch the air above the clatter of cups and saucers, and trays being emptied behind the screens by the table clearers. The group's three coaches, all men, sit away from their charges.

They are joined by a rugby team with 'England Rugby XV' on their backs: club ties of blue and red on large men with Popeye muscles, big arses and baseball caps. One wears shorts and trainers. A giant prop forward has a couple of cauliflower ears hanging from his head.

Outside, the blue sky filters through the aircraft hangar roof of the building. An endless flow of people enters and goes off in the direction of the toilets.

A couple of grey-haired old ladies sit next to me. I am smiled at. That pitiful look that people give you when you're sitting alone in a restaurant. Their dentures talk incessantly as they butter their toast and sip their teas.

A father has his young son sitting on his lap, then places him on the table. The young child looks around in amazement. He coos like a pigeon and is sick down his front.

A young woman in glasses, black hair tied up behind, black jumper, painted fingernails with stars on, stares at me whilst munching fast with her mouth open.

Two full breakfasts sit in front of me: eggs, tomatoes, fried bread, bacon, sausages and beans. The woman is dressed in a mauve top and cheap jeans and looks as if she has just emerged from a shower; her dark hair hangs limply over her breakfast. Her man, with a Robbie Fowler nose, eats impassively. They avoid eye contact. The occasional word slips out between gulps of food, and crosses the table. Both are wearing dirty white trainers.

A nineteen-year-old, in glasses, wearing a jumper with no shirt, blue jeans and monkey boots, rocks to and fro munching a green apple. One arm is folded around him.

Two table clearers, both male, exchange a few pleasantries as they move between the tables; their East European accents turn a few heads. I read 'asylum seeker' in one person's face.

People are walking around looking at the sights. A baby is crawling on the carpeted floor. His father watches before picking him up and swinging him. Groups of people huddle in the entrance concourse. Is this service station the only place for human contact for miles and miles of soulless motorway?

The Grand Prix outlet has a sale but the shop remains empty apart from a man in a French policeman's hat. He ignores my approach.

Decide Saturday is a trainers and jeans day!

8 January

Roseanne is serving by herself. Her thick Scottish accent baffles the early morning customers.

'Are you being served, love?'

My two £1 coins disappear into her plump white hands.

A group of Chinese young people sip Cokes whilst others play on the shooting range in the Game Zone. They are shepherded by a tall Englishman in a brown suede jacket and an expensive checked shirt.

A young family hide in the corner. Their three children, all with blond hair, bubble around in their seats. Dad is bald-headed, like one of the Mitchell brothers out of EastEnders. Mum is a chubby woman, her fawn-coloured top stretches to hide her full English breakfast.

I face three fleeces: one in off-red with his breakfast tray empty alongside him; a dark-blue fleece, black shoes gleaming; and a Vodafone fleece; with a woman with a grey-hair bun.

Four thirty-something dark-suited businessmen sit at a table next to me drinking tea and toasted La Brioche Dorée egg and bacon sandwiches. Talk of 'Location maps', 'extended details', and 'Steve's fault'.

10 January

'Are you waiting?' I ask a balding man dressed in a knee-length blue rainproof with a well-travelled rucksack on his back.

'I am invisible,' he replies.

June serves him.

'Yes, sir?'

'Do you have croissants?'

June blinks, not quite understanding him.

She moves her large body over to the serving hatch.

'Could we have some more croissants, please?'

From where I stand all I see through the hole in the wall are stainless steel units, but no people. A woman emerges from the kitchen carrying a wicker tray and on it, ten croissants.

'How many would you like, sir?'

'How much are they?' asks the invisible man.

'99p.'

'I'll buy one please and may come back for another.'

June smiles at me as he moves up the counter to pay.

I notice he has a Qantas logo emblazoned on a credit card as he nervously fingers a £5 note from his wallet.

He retreats to a table near to the entrance, and cuts a lone figure as he consults a large map. Is he walking around the M25? I ask myself.

A family of three sit alongside me. Two sixty-year-olds and their spinsterish thirty-year-old daughter. They are en route to Cheltenham. Toast and tea is consumed in total silence. He wears a green anorak, brown sensible shoes from Clarks, a check shirt and functional nondescript trousers. His wife looks as though she has dressed for church. Comfortable green coat, still buttoned up, with a large brooch pinned to one lapel, matching earrings and black 'normal' shoes. Their daughter has that physiotherapist-at-rest image, white Fred Perry shirt under a non-designer fleece. The two adults wear high street glasses. The cups and saucers and teapots are neatly stacked on to one tray, serviettes are folded on one side. I notice the man wears one of those funny insignia rings, which identifies him as a 'secret hand-shaker'. The three toddle out; the daughter's red and black Gore-tex jacket hangs off her shoulders. Her trainers squeak as she walks across the floor.

11 January

I park facing the travellers' site. They have disappeared overnight.

Red gas bottles remain along with small piles of rubbish. A shop model's leg pokes out of one of the pile of boxes next to a bundle of foul-smelling clothes. Two men in Welcome Break zoot suits are cleaning up the site. Black plastic bin-liners are stacked on to a luggage transporter which looks like a milk float from a previous life.

Sandra is serving in the shop – a white, Englishwoman in her forties, with her glasses hanging from a strap. She wishes me a good day. I return the hope.

Some out-of-towners – a late middle-aged couple. He wears golf-style check trousers which flap around his ankles, and light-coloured socks with brown walking-type shoes which have been polished beyond their time. Hair grows out of his ears and nose. He has a furrowed brow, and bushy eyebrows which give him a Dennis Healey look. His wife wears the Marks and Sparks shoes which only grandmothers wear. Her blue trousers match her jacket. Overweight, she trundles alongside her husband as they both bounce on their feet en route to the toilets.

A coachload of young businessmen and businesswomen appears. Suddenly the place is full of Next suits and accessories, bit like a Mormon convention. The shop is brought to a standstill as bottles of Oasis are purchased. Their deep red colour seems to be the 'in' drink.

'Where's the toilet?' a young woman in the group asks.

The table clearers seem to be hiding. My table still has the previous occupant's teapot, and empty raspberry jam cartons. Where are they? None in sight! Most unusual.

16 January

'Beans or mushrooms, Tony?'

'Beans, please'

Two businesswomen sit across from two female students – one pair dressed in black trouser suits and matching black briefcases, the other pair have that post-Christmas essay look. One has her feet up on a chair, is intense, and sips bottled water. The other one has a Mediterranean look, hair brushed back, rather unkempt, black jumper and white jeans. They eat out of white anonymous plastic bags.

'Another day in Paradise', by Phil Collins, is followed by 'Nights in White Satin', from the Moody Blues. Secretly I tap my feet beneath the table.

The shop fitters are in: five short men, all Scousers, middle-aged with a young-looking apprentice, all wearing Adidas trainers and baggy tracksuit bottoms. Bums and bellies poke out. The younger member of the group is in animated conversation on his mobile phone – 'Don't worry love… yeh… yeh… ta, love.'

They embark on a shopping expedition. Cassettes and CDs are turned over in the square bucket which advertises 'Bargains'.

'Seen the fucking price of that, Jack?'

'Jesus.'

One of the group is picking his teeth with a pen knife after his Fat Boy breakfast as he ambles around the shop fingering the passport covers and gaudy mobile phone covers. Relaxed, at peace with the world, he looks almost innocent. The five of them disappear out the doors. I catch them driving out of the car park in an unmarked white transit van. Sounds of laughter and false farting come from within. An empty plastic drink bottle flies out of the window as the van is swallowed up in the M25 traffic.

17 January

Judith is a GP who lives nearby with her family. She walks to the service station from her house on a path that was the old London to Holyhead coaching road that now passes under the motorway.

'Sometimes I come in with really muddy boots and I feel the need to take them off because we have walked through the back and people might think where on earth have these people come from in Wellington boots. It just seems to me that that isn't acceptable. So I take my boots off and walk through to buy something holding my Wellingtons. I don't know if that tells you anything about the place or more about me.

'I come in for newspapers, and sometimes we get a Burger King meal. If I run out of milk I might come down here. The children get sweets from the shop. I wouldn't let any of them come here until they were about thirteen. I wouldn't let them go on their own when they were younger because they could have gone anywhere in the country from that point and I wouldn't have a clue where they were. So I made a rule that they weren't allowed to go to the service station unless there were the three of them.

'Sometimes they come down here in the holidays because there is nowhere else for them to walk to here. There is no other shop within walking distance. It's our corner shop.'

*

Welcome Break is the UK's largest operator of motorway service stations and attracts over 70 million visitors annually across the UK, which makes its stations a bigger draw than any individual UK shopping centre or airport.

*

20 January

It's Seventies morning. 'Hey Mr Tambourine Man', by the Birds gets me into a nostalgia trip as I sing alone over their business and social chatter. 'You're the kind of guy', and some other Motown classics follow. I am in heaven.

Three people are sitting at the table in front of me as Paul McCartney is singing the 'Long and winding road'. A black guy in a Hawkshead sweater sits facing a young white woman wearing a Levi's denim jacket. They are not very talkative. He chomps his breakfast. Facing them, is a kindly faced granny, in her red coat with black lapels, a freshly pressed white blouse with brooch, and grey permed hair and over-sixties' spectacles. I am sure she is tapping her feet to the Beatles' song in her Marks and Sparks fawn coloured shoes, with laces which join together in the middle and leave the toes exposed. She talks at the young woman and around but not at the man.

She produces an airline ticket and leaves the tables, her slip-on shoes crunching over the purple haze carpet. The old woman's pleated skirt hovers above her ankles. She is clock-watching; what time is the flight? Heathrow, or Gatwick? The M25 can take you in both directions.

A group of people down from the mountains enter. Overdressed ramblers? It's difficult to tell.

The Saturday staff are making busy everywhere. Ruud Hullit swings a mop on the main concourse floor in between balancing a football on his shoulder.

Arm in arm, a middle-aged couple swing out. Are they off to the Lake District or just in between shopping centres?

'Cathy's Clown', the Everly Brothers' 1950s hit serenades me as I head off to the toilet for a No. 2.

'The time?'

Ruud is now the toilet cleaner/attendant.

'It's about ten o'clock,' I reply.

His head bends towards my wrist to see the time. Not sure he understands. Leaving me he writes the time up on the wall chart to say he has inspected and cleaned the toilets on the regulated hour.

I sit peacefully in cubicle 4.

A mop-head appears under the door narrowly missing my *Guardian* 'Review' section which is lying on the floor at my feet. The mop disappears and I hear a gasp from the next door cubicle.

Leaving.

Four Hell's Angels stand around the entrance smoking; men in their thirties, large, black-booted, one the size of Big Daddy, the Saturday afternoon TV wrestling favourite from years ago.

They're German, of course; the spiked World War I helmet on one of their heads gives the game away. Newly arrived customers stare and avoid walking too close to them. Mothers pick up and carry their small children past them.

With their black hair (one has a shaved head with a ponytail), silver studs on well-worn black jackets, and thigh-clinging greasy leather trousers, they move slowly in convoy towards the toilets. One woman crosses herself as they pass her en route to the gents'.

*

Welcome Break seeks advertisers in the bathroom facilities and can increase revenue depending on actual numbers of people potentially seeing the advertising.

*

21 January

The travellers are back. Eighteen caravans have appeared and now sit on the frost-covered car park area where they have set up camp. The familiar tall red gas bottles lay scattered around the caravans, some in use, others discarded. Second-hand Ford cars are parked up haphazardly. There are few signs of life from the encampment. No dogs! Curtains drawn, blinds down, a few white transit vans, one with the London Building Company emblazoned on its side. It's Sunday morning. Wet. The rest of the car park is empty.

Inside, an African security guard talks to a customer in French. A lone player sits in the Game Zone. Four burly traffic cops sit together with their full breakfasts. A couple of family groups take up two tables, the men and boys sit on one table, and women and small girls sit on another. They are passing through.

Lillie, young, hair up behind her head serves me my chocolate croissant and a pot of coffee. She smiles as she short-changes me. I let it go.

José, the table clearer, flicks his blue cloth from hand to hand in expectation as he patrols the tables.

Lots of 'Northerners' on tour; strange accents rise above the Thames Estuary dross.

A balding man in a blue sweater over a light grey shirt, smart black trousers, and black jacket on the back of his chair, sips his coffee quietly and effortlessly.

'Junio!' He has shouted at the top of his voice for no apparent reason.

Heads turn. One of the policemen eyes him suspiciously. Junio does not appear. We return to talking, reading our newspapers, staring into space.

'Bitches!' is shouted.

The 'quiet man' sips his coffee. No emotion, no discerning difference in his behaviour or facial expressions between sipping his coffee and shouting out!

The policemen turn around and look at me accusingly.

I leave.

22 January

A mother and her teenage daughter sit face to face. There is no conversation. The daughter looks embarrassed as Mum sips her tea, one hand on her chin with her left elbow on the table. She smiles and gives a 'no' shake of the head as Mum talks to her. They look marooned in a sea of empty tables and chairs. Mum has her handbag over her right shoulder as if guarding it.

Two men; one white, a dead ringer for an off-duty policeman in his blue shirt, dark-blue serge trousers and black shoes talks across the table to a white-shirted black male with dreadlocks tied behind his head in a ponytail, his suit jacket off. The 'Old Bill' meets who? 'Old Bill' does all the talking: demanding gestures as his feet tap the floor beneath the table. They leave separately.

Later...

The place is 'buzzing'. The car park is full, and most of the tables are taken. All of humanity is present. Newcomers enter and check out the interior, the décor, before agreeing to proceed, like checking out a person's interior design of their flat before jumping into bed with them.

A young Karl Marx lookalike with blue suit and waistcoat, fresh from the British Library, walks across my vision. I am sure he is carrying the first draft of volume one of *Das Kapital* or is it *The German Ideology*? Nigel, out of the *Young Ones*, sits with an older 'hippy' with a goatee beard and a young woman and baby. Four policemen, two with flak jackets, sit in a huddle nearby. Business meetings are going off all around me.

Two fat people, a man and a woman, eat a bag of doughnuts. She wears an extra large size tee shirt with 'Exotic' emblazoned on its front and has an earring in her left ear. The man has close-cropped ginger hair. Tattoos run up and down his arms. His mobile phone rings. He grasps it with one of his big paws. They waddle off together. Doughnut crumbs lie on the carpet beneath the table.

Two more fluorescent yellow-jacketed policemen join their colleagues sitting at the 'police officers' table'. Quite a gathering. Talk is of shifts, one particular arrest last night and 'male' banter.

In the car park I am surrounded by other VW cars: a Passat, a Bora and a couple of Golfs. Two Rovers nestle alongside. A cluster of Vauxhalls stands nearby. An unsmiling Sadaam Hussein's twin squeezes into his VW alongside me.

29 January

'Thought I better tell you, sir, that the price of chocolate croissants is going up by 45p tomorrow.'

'That's a lot.'

'I know. Scones are going up by 50p.'

'How about coffee?'

'Don't know yet, sir. As one of our regular customers I thought I should tell you.'

'Thanks.'

'One customer comes in each morning and buys three pains au chocolat… That's £3.29, sir. Have a nice day.'

A Welcome Break 'Big Boss' in his long expensive dark-blue

trench coat, glasses hanging down his front, is showing three young wannabe managers around the site. All three look eager and keen to please – two female, one male. They have that earnest graduate look about them. Two of them take copious notes as the Big Boss emphasises a point with a wave of his hands. Staff appear to be on 'red alert'. Even the coffee outlet has a cashier complete with Welcome Break cap on duty. The table clearers are working overtime.

An unusual sight. A black male businessman in a dark-brown suit clutching his mobile phone in one hand and keys in the other is introducing himself to an Asian businesswoman. They shake hands. He sits away from the table in defensive mode. She leans forward, spearing her fruit salad breakfast. She is expressive, and articulates using her fingers. Her spoon is used to good effect. He's out of his depth; if he reclines any further he will crash to the floor. A camel-coloured coat covers her white blouse. A dark grape is slipped between her red lips. He looks up at the ceiling as he makes his point. She flicks her hair. He fumbles with his phone. His watch comes off and is put on again. He looks ahead and walks off. Her coffee is delicately poured.

30 January

Chris serves me my chocolate croissant and pot of coffee. His blond skinhead No. 4 haircut gives him a menacing persona.

It's a Seventies morning with a Dire Straits record followed by the strains of Van Morrison's 'Have I told you lately that I love you?'

Into the gents' for a No. 2. Need to write up some notes in my diary here in cubicle 25. Sitting, my pen slips out of my hand and rolls under next door's cubicle wall. I sit transfixed. I can hardly put my hand under the floor to wall space and grope around for my pen. I sit. A hand appears from under the partition walls, attached to a blue suited arm, my pen is in his hand. I take it back and mutter something incomprehensible like 'Sorry about that; thanks.'

I quickly leave but wash the pen as well as my hands – just in case!

31 January

In early. Sleep in my eyes. Get my *Guardian* from the shop and almost walk over a small man who's less than three feet tall, carrying two bottles of orange Fanta. His long trousers flap over his shoes. He peers over the counter at Narine who takes his bottles.

'That's £1.98, please.'

'But it's only two bottles of Fanta,' he protests.

Narine says nothing and looks at the man's small baby hands.

A credit card is produced. The queue moves over to the other cashier. He is left by himself as the credit card sucker machine checks his card. I leave, looking behind me to see him reaching up to the counter signing the credit card slip.

The late ex-President Nasser of Egypt is in this morning; tall, a very well-built man, wearing a black leather jacket. He has short, closely cropped grey hair. His compatriot wears a blue and red winter jacket. A large broken nose juts out from his receding hairline. One hand on the side of his face, elbow on table, he grimaces in pain. He changes hands, covers his eyes and forehead with his other hand. He looks ill.

With his blue jeans and desert boots, hair greying at the sides, he looks like an East End villain straight out of one of the many books written about the Kray twins. He seems down on his luck. In his red scarf and blue denim shirt open at the top two buttons, he walks off to the smoking section with his cup of coffee. I notice his laces are undone and his jeans rest on the top of his boots. He has that familiar ex-prisoner gait.

A whistle from him attracts the attention of ex-President Nasser who has ambled back from the toilet. His dark-blue shirt and black and white diamond tie covers his protruding belly and undone fly. They disappear into the early morning fog.

I listen to Richard. He's a hard selling sales representative area manager. His fingernails are bitten back to the quick. His once dark hair barely covers a growing bald patch on top. A black suit is matched with a yellow checked shirt and blue-flecked tie. He says he is in his late thirties but looks much older. Andrew sits across from him in a bottle-green suit. Andrew's the apprentice. Richard does the talking. It's short. Snappy. Words are fired off.

'Good place to meet.'
'Convenient.'
His mobile sings.
'You must sell.'
'Tony will have to go with it.'
'Look at it this way.'
'Every time we gave it a bit more it gets there.'
'What did we get back?'
'Staff like being told.'
'What?'
Andrew tries to respond.

He listens but looks away. His black case is packed away and he's on his feet.

'We want those sales... supermarkets... key,' he talks over the man.

'Be very, very careful.'
'Sorry must go.'

Both stand up. The mobile phone conversation continues. The bottle-green suit follows the black suit to the exit.

11 February

James is a sales assistant in the shop. He has red cheeks and slicked back hair like a ventriloquist's dummy. He's pleasant to customers. Some even queue at his till for him to serve them because he is so nice. His blue work Fred Perry shirt bulges in the middle and hangs over his black trousers. I watch him delicately wrap two bunches of flowers for a middle-aged woman who is in a rush but has chosen his till. A queue begins to form behind her. Feet begin to shuffle. A few tuts can be heard. Menacing glances from impatient people waiting.

The shop is busy. Ben Sherman shirts have been reduced to £19.99 which has caused quite a stir. Green felt toy dogs are pawed over by old ladies carrying tuna fish and cucumber sandwiches. A little girl with her father requires precisely three lollies from the sweetie bargain basket.

Two coachloads of 'Brummie' businessmen enter – first-time visitors. They join the queues at the wrong end and collect their cups and saucers first before ordering their teas and coffees. This throws the system into chaos.

Two Asian members of the group sit alone at a table; one a Sikh, the other a Muslim. They both wear dark suits. Everyone knows each other. It's a work's outing to London. A few women are sprinkled amongst the laughing men. They also sit alone.

'I am still looking for Sandra,' explains one.

Mobile phones ring incessantly.

Sandra has appeared to cheers.

The noise increases to Concorde lift-off levels; it's difficult to hear the in-house music. Men with trays full of coffee pots and croissants peruse the tables. It's beginning to feel like a gay pick-up bar. The 'girls' are now giggling at their table. Male and female hormones are flying through the air.

Percy Sledge's song, 'When a man loves a woman', drifts across the aircraft hangar. It's just audible over the sounds of dishes being washed and hung out to dry. It's good. I gaze at the sun glistening on the metallic tables outside. Blue sky is playing hide and seek behind leafless trees and posts outside.

'Hey, you from the Dudley coach party?'

15 February

An unusual sight, some of the tables are cluttered with trays.

See the tall lanky youth and his younger brother again. Two large hearing aids are attached behind his ears. His teenage blond hair is unable to hide them. He saunters along behind his older brother, his face hidden by a blue baseball cap, as he leads them out through the sliding doors exit.

Chris Bonington and his mum have just walked in. She is clinging on to his arm. Chris looks out of place in a grey lounge suit as he stumbles around to the toilets. His mum's long pink woollen scarf and ankle-length coat, remind me of a female Dr Who, aka Tom Baker.

Passing through the crowd. The cashier at the shop checkout says hello to me as I purchase a paper followed by a goodbye. At the 'Coffee Pot' conveyor belt Michael asks how I am. I looked dazed and reply, 'Fine thanks, how are you?'

'Used to see you a lot when I was working in the shop, you used to come in there a lot,' he informs me.

I am sure I didn't but I smile and wish him a good day.

Two men are playing backgammon at a table, complete with foldaway case, red and yellow counters, and pot to shake the dice. Discover they are French, and they have that particular French trait of looking good even in the most ordinary of clothes.

Two middle-aged men, with closely cropped hair, look a bit like ex-jailbirds, in casual gear, one wearing an Adidas tracksuit and CAT trainers, the other in a Kicker fleece. Both have earrings in. They look slightly disorientated.

A 'lost' coach rider, a pensioner in his seventies, walks from queue to queue hoping to pay for his two slices of toast and pot of tea. The counter staff ignore him. He walks off without paying.

A large Asian family sit next to me, all talking at once. Several children of all ages excitedly chat away as if they were going on holiday. They have impeccable middle-class English accents.

Outside in the car park a white Ford transit carrier has four men standing around smoking fags. One geezer stands to the side whilst the driver sits patiently drawing on his roll-up. White smoke like early morning clouds hover above them. The others are moving gear from the hotel where they stayed the night. They are dressed in blues and black, and look reminiscent of the infamous Black and Tans. Talk is of last night.

'I tell you she was thirty-five if she was a day.'

'Bollocks.'

Much laughter.

'You're joking.'

'You had seven pints I'm telling you.'

They are 'shop fitters' on a job a long way from home.

21 February

I have seen the 'Fat Controller'; his grey hair swept back, a portly figure, resplendent in his Welcome Break uniform of blue shirt, yellow speckled tie, and dark-blue trousers. He is patrolling the 'decks'. The Reverend Awdry would be proud of him.

Two young boy soldiers with No. 1 haircuts wearing army fatigues tuck into their canteen-style breakfasts. An occasional word passes between the two. One has finely chiselled features – would probably pass for a ballet dancer in another life. He is playing a form of 'footsie' under the table with his mate. The latter has a 1000-metre stare. Maybe he's seen action somewhere – Kosovo? Catterick Army Barracks? Pig and Whistle pub?

A blond woman wearing Cuban-style boots with tall heels sits at one of the 'singles' table, delicately spooning her cereal. A glass of orange juice waits alongside her. Coffee is poured. Her long, slender, artistic fingers hold the pot with care. A man, with bulging cheeks full of sausages and bacon, gawks at her undertaking this manoeuvre. He catches sight of me looking at him – embarrassment, like being caught by your mother coming away with a 'top shelf' adults' magazine from the corner newsagent.

The Fat Controller re-enters the hangar carrying a yellow traffic cone. His belly bumps into the cone as he walks along carrying it in front of him.

Men's eyes follow the blond woman as she rises from her chair. Her black handbag placed on her shoulder, a swift, dancer-like movement, and she is walking past me. Her expensive perfume kills the smell of fried breakfasts. Her glasses reflect wasted male gazes. She is in her late thirties, early forties. Her catwalk ends via the direction of the shop. Suited businessmen crane their necks to catch a momentary glimpse of her. Like African Suricates, or Meercats as we commonly know them; some stand on their back legs to get a better view of her.

I drink my tepid coffee and escape into the cold morning air to cool down.

*

All you need can be found at a Welcome Break. When you need a break on the motorway, our welcome is the warmest you'll receive. If you want a meal, a newspaper, a bed for the night, a full tank of petrol, or simply to stretch your legs – a Welcome Break has all you need.

*

22 February

A hive of activity. Business meetings all around. Mobile phones ringing in the day.

Catch a whiff of cheap perfume as someone tucks into a

chicken burger. Manic children running around (is it half-term?). People in transit wandering around not knowing the procedures for using the service station, joining the queue at the wrong end, inability to find the toilets, walking against the flow rather than with it!

People are shopping in between boring motorway journeys.

'Oh look, two for £2.69.'

'It's a bargain.'

'Debbie would like that.'

A middle-aged woman with her husband, both very suntanned, more Marbella than Morecambe, hesitate with their purchase.

The smell of Daddy's Sauce woofs across my table. Eggs lightly done come next followed by a burp from behind.

Three businesswomen to my left, in their twenties, all dressed in varying hues of black – two ponytails and a bob cut. Paper swishes across their table. A laptop computer whirls to a delicate touch. Much animated discussion punctuated by silences as three pairs of eyes gaze at the computer screen. Assorted bags litter the floor around them.

I have been in Hackney this morning, having my 'usual' Friday morning breakfast of two eggs on toast washed down with three mugs of strong, Yorkshire Tea strength tea. Spent the entire time since then back and forwards to the toilet.

Get caught short again; I dash to the gents'. Only just make it. Washing my hands I witness a white-haired man, in his late seventies, meticulously cleaning something in one of the sinks. Soap bubbles everywhere. Something is stuck in the plughole. I catch sight of a denture grinning at me through the bubbles.

3: 'I think he might be Lord Lucan'

You must learn the lesson fast and learn it well
This ain't no upwardly mobile freeway
Oh no, this is the road
Said this is the road
This is the road to hell.

Chris Rea, The Road to Hell (Part Two)

Carol feels like a skivvy – runaways? – Dave says Good morning – it's becoming an obsession – bikers off to Kent – Pauline patrols the decks – Mum asks where she is – no Comic Relief card for me – my Doc Martins get a rinse – Princess Diana.

23 February

Little old lady, sitting alone in the corner by the entrance, eating a small blueberry muffin. Her woollen tea cosy hat is pulled down covering her forehead. We talk.
 'My son drives me here every fortnight.'
 'I like to come here and watch people.'
 'Made a few friends here.'
 'It's a bit expensive but I like the view.'
 She sits facing the entrance. People ignore her. She surveys the scene in between her false teeth nibbling bits of the cake as it sits within her wizened fingers. I buy her a pot of tea.

24 February

Carol is in her twenties, and lives locally. She has worked at the service station some four years. We sit and drink coffee as she talks.
 'We are called sales assistants but it feels like being a skivvy

sometimes. It changes every day, cleaner one day, cook the next, cashier another day.

'It's all shift work. 6.30 a.m. to 2.30 p.m., or 2.30 p.m. to 10.30 p.m. You get a few who do a middle shift in between, like 8 to 4, 10 to 6, but those are the two main shifts. It depends whether the manager likes you or you have done him a favour. If you have done him a favour you get some nice 8 to 4s and 10 to 6s. If you haven't, if you say 'No, I can't do that shift,' you get all earlies for a week or all lates.

'It's stressful sometimes. It depends how busy it is. We are serving the worst public basically because they are travelling on the roads, stressed out. They come in here for a break and then they are hit with awful prices and we get it.

'They don't seem to understand that we don't make the prices. They have had a nightmare on the motorway, been sitting in traffic and they take it out on us.

'It can be good fun too. In a place like this you can see all sorts of people. We get coachloads of old people, we get families, we get groups of young people – we had nuns in today – we get the police coming in. We get prostitutes. You can tell. You can read people. We get gypsies. They come up, order their meal, pay for it, go back, sit down, eat half of it, wait half and hour and bring it back for a refund.

'They got away with it for a while but we have knuckled down on it now. They know every trick in the book but they don't realise we are onto it.

'Coach drivers are the absolute nightmare. We have had multiple arguments with coach drivers. They come in and think they own the place. They walk in to the front of the queue. They are getting a free meal and they expect to be served first. It's one meal, one drink, and one dessert. We have some with three or four drinks on their tray thinking they can get it on one card for a penny, because they can get a meal for a penny if they have got a card. They all forget their card. Every other driver has not got a card. Because it's only one visit a day we think they do it on purpose so that they come in later again. I caught one out because I did a double shift and he was in in the morning and late at night and he wouldn't believe I was still working!

'I haven't got my card, I have lost my wallet. Everyone seems to lose their wallet, every driver does.

'It gives them power, they think we are their little skivvies. They click their fingers and expect us to go running. You get their meal and they go "Get a coffee then", "Get me this", and you think, what are you talking about? There have been mega arguments with them because we won't put up with it.

'The best customers are businessmen and the police. The police are so good. They get a discounted meal and what is good about the police is that usually in your life you only speak to police when you are in trouble. But here they have a joke, they tell us stories about what they have been doing and they are better educated people obviously than the coach drivers and they know how to act. They are in a powerful job; they don't need to exert power and try to use us. That is the problem with the drivers, it is the only bit of power they get in their lives.

'Old people are funny because they are in another world half the time. "Where are the cups dear?" and there is a big cup stand in the middle. We must tell, on one shift, fifty to a hundred people where the cups are.

'In The Granary Express they come up, they pick up the soup bowl because they can't find the cup and think it is a cup, put a tea bag in it and put it on the hot chocolate machine and fill it up with hot chocolate, so you have got a soup bowl with hot chocolate and a tea bag in it and then they ask us what is going on.

'So usually we have to get someone to stand by the self-service machine. On the machine there is a button that just says "press for hot water", "press for coffee" and they can't work that out.

'We have some regular customers. They come in with their elderly father on a Saturday at half one. He likes coming here and he can't get out a lot. They come in with a big bag of change. They collect their change each week and use it for meals. They bring backgammon, snakes and ladders, and scrabble, and play board games.

'They are tiny little people, really small, really short people. He pretends he is new. He asks where the cups are and how much is the coffee. They're a Jewish family from North London. He is a doctor of some sort, I saw his credit card once and it said

Dr whatever his name is, but you wouldn't think it to look at him – flyaway hair, shirt hanging out. They stay until seven at night. They have a Granary meal, coffees and teas, then they will have a Burger King and then more coffee. They buy the magazines from the shop, take the free leaflets out like the scratch cards, throw the magazines away and play on the scratch cards.

'We have the fish and chip people every night. They got a bit upset because we changed to haddock. The cod changed to haddock and they got very, very upset. It's a man with his mother; I am assuming the mother is in a home and he picks her up and takes her for a meal here but I don't know for sure. He comes in and orders fish and chips for her and a cappuccino but she acts like a child.

'People meet here. I have seen people having relationships meet up, and sit in the corner together.

'We had one couple that were living in the hotel at one time and they used to come in and there was something dodgy going on. They were living in the hotel for over a year. We found out through other people that they used to bring a bin bag of things from their room and put it in the car every morning. We think they used to go down to the ferry, go to France get cheap booze and things and sell it.

'They were a dodgy-looking couple anyway, and it was obvious something was going on there. They didn't work because they were often in and out at funny times but they always had a lot of money.

'We have professional complainers. This couple, a grown-up middle-aged couple, they used to come in and complain. They used to have one meal and share it between two and because I knew who they were I started plating it up how I knew they had it and they were quite pleased with that! It was funny but then one day they came in and they went absolutely mad because somebody was in their seat. Like it has their name on it. A seat in the service station! You expect a child to say that but not middle-aged people.

'I had one boy the other day; I was standing behind the counter working on the till, full uniform, hat, everything. 'Do you work here?' and I said, 'No, I am just standing in uniform behind the till for fun.'

'I have seen people passing things. It's an easy meeting place for them. There was £1,000 found once.

'We get things left behind as well; toys, dolls, and things, bags, keys. It amazes me how people walk off and leave their handbag on a table. Someone left one and she got me to send it in a cab all the way up the M1; it must have cost her about £50.

'A lot of lonely people come in here. They come in on their own. They sit for quite a while with a coffee and just sit around and watch. It's good for people-watching. I do it myself! We stand behind the till and if it's boring on your shift, you watch people and it's just amazing to see what people come in.

'Weekends are the busiest times. Weekdays like Tuesday, Wednesday and Thursday are the quietest days, especially the evenings, but Mondays are coaches, Fridays are coaches and the weekends mainly because of the coaches again. We get a lot of army, TA at the weekends.

'I have had phone numbers given to me. I had one the other day who wanted me to come up to their hotel for the night, they were businessmen. I didn't go. There were two of them sharing a room. I am not that desperate! I have had phone numbers from builders; some people come in for things like that.

'I have done this job four years on and off doing part-time, full-time, and holidays. I have just finished a degree, throughout the time. So I am going into hotel management soon.'

26 February

Late evening.

The car park is empty apart from a couple of gay men cruising. Inside is equally deserted. Four people playing in the Game Zone. A young couple both wearing glasses stare at their separate machines. Coins are continuously fed into hungry metallic mouths. The man's baseball cap is too big for him. She has a dishevelled look.

Two younger men sit facing their own machines, one has blond short hair, shaved around the ears. His small hands pump up and down. His friend has jeans with six-inch turn-ups.

Outside two thirteen-year-olds sit together deep in conversation. The boy with an Adidas jacket, and a Jewish skullcap, fondly puts the girl's white scarf in place around her neck. She looks young for her years. Her desperate eyes watch me as I walk past. Runaways?

27 February

The young teenager with the hearing aids is in with his mother. They have a routine. They sit in their usual place, the stool and table area, where he sits with his older brother. He talks incessantly. Loudly he asks his mother what she did when she was at school.

Danny at the Café Primo says Good morning to me and asks me how I am. Does he know me?

28 February

I'm beginning to get an unhealthy obsession with this place. I find myself lying, making up excuses, sneaking off work early, and creating appointments with fictitious people so I can spend time here.

1 March

Tension at the coffee counter. A small man, going bald, in a blue ski jacket, with a mobile phone caressing his ear gets overlooked by Tony.

Another man asks if this is a queue.

'It's like it all the time, even at 4 a.m. in the morning.'

Abdullah serves me my coffee and croissant. 'Enjoy your coffee and croissant and have a nice day.'

A coachload of sixth formers appear, all very polite and civilised. Young boys and girls read their *Sun* newspapers. Probably the only time they get the opportunity to do this is when they are away from their middle-class *Guardian*-reading parents. Mixed groups sit together tentatively negotiating boundaries. Some have spotty faces, one or two wear baseball caps. One goes for the druggie look wearing a 'Rage against Nothing' emblazoned across the back of his hooded sweatshirt.

The rat-a-tat of the gun at Game Zone rises above the chatter. A young looking seventeen-year-old in a beige jacket, decapitates androids with an Arnold Schwarzenegger Terminator long-barrelled pistol. A group of four of his friends moves to the rifle destruction game. One takes aim and fires. The others watch, spellbound, as bullets fly into bodies.

Burger King remains closed; its lights dimmed, colours hidden, cash tills silent. A few expectant customers look confused. Has it been shot by mistake?

3 March

Friday evening.

The Burger King outlet is open. Chips are everywhere. It's quiet, apart from the gaming machines busy offloading noises. A group of young white men in baseball caps and jeans stand watching transfixed. The car park darkness outside ups the lights in the Game Zone.

A couple of middle-aged men do a gay step dance and depart in different directions – one to the toilets, the other for a cigarette outside.

4 March

Two middle-aged bikers, in black leathers with red trimmings, and Nazi-style black knee-length boots, talk of bikes and women. They sit around drinking Diet Cokes.

I join them.
'We're off to Kent, south coast, to visit some chicks.'
Laughter.
'His woman, not mine.'
'Go on, you would, given half the chance.'
'It'll take us just over an hour to get there.'
'We do this as often as possible.'
'Stop here for the toilets and a drink.'
'Mind you there's some good crumpet here.'

Coaches have disembarked a Saturday Northern crowd. John Lennon's 'It's over' drifts like a pall of cigarette smoke across the twang of Liverpudlian accents.

5 March

Pauline has worked at Welcome Break for eleven years. She patrols the concourse as a troubleshooter. Previously she worked in a restaurant at the BBC studio in Borehamwood where they make EastEnders. She also did security work for them. There is a Pat Butcher look about her. She lives locally.

'I'm Customer Services. I did five years in the shop as cashier, then went on to Customer Services. I have got responsibility for the Game Zone there. So I liaise with engineers, cleaners, and my boss so that everything is kept working and operational, and making the money they are supposed to.

'On a really busy day I make sure the toilets are clean, the ladies', I hasten to add, and the disabled, and the baby room.

'I'm sort of like a general gofer. I mop up all round the edges of what is not someone else's remit. It's a case of we need this done and so Pauline does. I quite like it because you have to think on your feet. There is a different situation every half an hour. You might call an ambulance one minute, do first aid, or you might be calling the police.

'That's another thing that I do. We have a coach drivers' club that we like the drivers to belong to. We can't hold a gun to their head and make them join, but we have got about 15,000 members so it can't be bad. Most of them will have their card, which is a swipe card for the till, and if they haven't got one I have got the application forms there I can photocopy their licence and get it all processed for them.

'They join because they will get at this moment in time limited discount on a couple of things. A coach driver gets a three-course meal for a penny anyway and some people would say for goodness sake isn't that enough. At this moment if they have a card they also get a pound voucher which is usable only in our shop, or one of our other sites, and they get discounts for staying at the hotel here. It's to encourage drivers to bring their coaches in so that the company gets the business from the passengers. So I try and get them signed up. The majority of them are no problem at all.

'It's like a big transit camp here. We are used as a meeting point, especially, I would say, for the separated parents. They will meet here and exchange children for a weekend and things like that. People have business meetings here. We have people coming in on coaches, coming back from holidays or meeting coaches to go on holidays. It's a big meeting place for people because we are literally the crossroads where south meets west and being at the very top of the motorway, it is just an ideal meeting place for people coming in all directions.

'You get quite a lot of famous people wandering through. Adam Faith was always a regular. I have probably not seen him in the last couple of years, but even when he wasn't on TV he used to pass through fairly regular, even if just coming into the loo.

'There are all sorts of things going on in the car park and you have only got to use your imagination if you know anything about people and the size of that car park. The CCTV cameras are quite good. Some of the things you don't want to be seeing anyway, specially when the second car pulls up and it's the wife who gets out and catches her husband in the other car with someone else. That happens regular.

'Some years ago one of the staff overheard a conversation between a couple of guys and as a result the police were called and the next minute we were surrounded. A couple of guys had loads of guns in the back of the car but it turned out that they were gun dealers but we weren't to know that at the time.

'We had a day, a couple of summers ago – we had the police helicopter above and this place was surrounded by the armed police. There was somebody in here that they wanted and that was all quite interesting, especially trying to get the public to move away. Two people who they knew would be with him were in the building and were taken out. It was quite strange because they had obviously had the place under observation because suddenly these people who you thought were drinking coffee out on the veranda were suddenly telling everybody to keep back and they were all the plain clothes bit.

'You get Customs and Excise following people up from Dover and confiscating vehicles and contents so the people left here are stranded to find their way home. You see it all.

'Mostly it's new people, people passing through. You do get to know your regulars; you get to the stage where you recognise a lot of them. Especially me. Watching the machines you get to know all the guys that come in and use all the machines and a couple of women that are in regular using the gaming machines. They're here every day or every other day. Some of them spend their weekends here. It's very sad but it's what they do. Most of them work, but a few don't and it's their only revenue. They are obviously good at it for the amount that they put into the machines, they get money out. It's an addiction. I know one young lad, all his student grant went into the machines and he was in a really bad way. Then he disappeared.

'You would be amazed at what people leave; passports, bag loads of money, valuable things that any normal person wouldn't leave. Mobiles come in three or four a day. I've seen the magazines that get found out in the car park and there are various other items, probably bought at Anne Summers' parties, left in rubbish bins. Whether they were actually used in the car park or not I couldn't say.

'We get truckers coming in quite often in the evenings; they prefer to come down here. As long as they leave the truck in the Truckstop I have got no problem. They do use our coach park later in the evening if they are full up over the road.

'People do complain. We have a standard letter that we send out to them to explain this and that. I actually say to them, if they come up and say 'My God, is this right that this cup of tea costs £1.69?' I say, no, actually it's a pot, it's two cups, it's not one cup and yet do you realise that we use 250 toilet rolls a week, 300 light bulbs a month. All these things 24-hour staff have to be paid for; unfortunately, it is certainly not in our wages! They base everything on Tesco prices.

'We had a call today that the travellers are circling. It's like cowboys and Indians sometimes. They go down past the Truckstop and have a look there and then they come round into our coach park and they get pushed out of there. Then they go round to the car park to the caravan area. We just track them. We have got regular families over a year and you get to know them and they're no problem at all. It seems to be the odd ones that seem to be going through every couple of weeks or if there is a local wedding they come out of the woodwork from all over the show. They will get in everywhere they can and unfortunately they are often the ones that cause the problems: their kids, their dogs make a mess, they steal stuff. They like the free showers of course.

'My worst customers are anybody who walks in to my Game Zone with a child under eighteen. They can't appreciate that the child is under eighteen so shouldn't be in there, so you usually get the "eff off" when you ask them to remove the child and they get stroppy. But you get used to it.

'Fellows in their twenties aren't very happy about being asked by a female to leave an area. The only thing is I think that because being female they don't start swinging their fists. If they start swearing at me, I say, "Do you talk to your mother like that?" I don't care if they are swearing and mumbling as long as they are going out.

'You might get another group of people who are just being a damn nuisance to everybody in general. You get a football crowd in and there is always one with the very large mouth and I just say, "Can you keep your voices down a bit because not everybody wants to be listening to you shouting and hollering." If they start singing, that's fine, no problem, always give them five or ten minutes and say "Right, you have had your sing song now, just quieten down."

'You get the verbal because you are asking people not to do what they want to do and they think they should. I have heard it all before so it's water off a duck's back.

'We have got seating for 450 people or so and our busiest times are probably Friday lunchtime, afternoon, Saturday and Sunday. So it's okay for the business people who aren't in then. Let's be honest – if you were meeting up to have a business meeting and you walked in and saw a load of the "Saga Louts", as we call them, all the little old people with their walking sticks, would you want to meet here?

'The old people don't know how to use the public phones so you are making phone calls for them, showing them how to use them. They find the public phones confusing. They stand in front of the telephones and ask you where the telephones are. Where are the toilets and there are huge signs everywhere. How do I get a cup of tea?

'The majority of people travelling around the M25 know where they are going, others don't and haven't even got a map in the car. They obviously don't read the signage because a lot of them arrive here and ask where they are. Like have you heard of Lakeside or have you heard of Bluewater? Just totally lost. You would think they were out for a stroll rather than a major journey on a motorway.

'People break down and a lot of them don't belong to the AA, RAC or anybody and you have to try and find the nearest garage and get them sorted. They just don't prepare for long journeys. They don't have maps, they don't have water if the weather is very hot. They are not prepared to be in traffic jams.

'People lose each other here. I get all the lost, sick and confused on the motorway. I've come to the conclusion that the entire public are strange. I really can't believe that some of them are let out on their own.'

*

'The travelling nurse says there are problems on the M25 Eastbound coming up towards the A1M interchange at South Mimms. Tail backs towards St Albans.' (Radio 5 Live Traffic News)

*

6 March

Cold frosty morning. A newish gold-coloured Vauxhall Cavalier is jacked up in the car park. Surprisingly all its wheels are intact. The driver's door and the boot are open. I go out of my way to have a look. The inside of the car is clean and empty. Where has the driver gone?

On the way in I pass a white Kenning Van Hire transit.

'Allo, Dave.'

'Alwight?'

'Yea.'

'What's up?

'Fuck off, buy your own fucking cigarettes.'

A group of men huddle together as they light up – some playful jostling; uniform of baseball caps, casual gear, tracksuit bottoms and trainers. One of the group remains in the van coughing his guts up.

A big green phlegm hits the tarmac. I am eyed with suspicion as I walk past. Two mobile phones go off.

'Give me a break.'

'Be there in thirty minutes.'

'Right.'

'Bollocks.'

I collect my coffee. No *Guardian* newspapers this morning so I slum it with the *Telegraph*.

Three men, with grey hair, easily into their seventies, with an ex-military air about them are taking centre stage.

'In the regiment punishments were never severe,' we are all reliably informed.

Red and black ties, blue blazers, and black shoes. They march off, one with his hands behind his back inspecting us civvies as they depart; grey flannel trousers, shoes highly polished, stiff backs, arms swinging by their sides. The three old soldiers continue talking as they walk away.

As I get up to leave, an elderly lady in a Thora Hird dress sits next to me with her daughter. 'Where are we now?' asks the old lady. Her coat is buttoned up tight to her neck, her eyes hidden behind cheap NHS frames.

'We're joining the A1M now, Mum,' her daughter replies.

The old woman looks around, bemused by the colours, activity, and the design of the aircraft hangar.

I smile back as they eye me up and down.

7 March

A member of staff has a quiet word with me.

'There's a grey-haired man hanging around the place at different times of the night and day. Someone said he's writing a book. Tony reckons he's homeless and sleeping rough. We get that here from time to time. Personally I think he might be Lord Lucan, you never know.'

8 March

A woman in her twenties with long blond hair and a Nicole Kidman face tucks into her full fried English breakfast. She is power dressing in a black business trouser suit, soft white blouse, and dark suede boots with large heels. A laptop is competing for space on the table along with two of toast and a pot of tea.

The Monkees classic 'I am a Believer' belts out as two male gorillas, in tracksuit bottoms and desert boots, manoeuvre their way out of the exit door.

I am beginning to feel left out as I am still not being asked by the two saleswomen sited by the toilet exit if I would be interested in a Comic Relief credit card.

As I walk past a white transit van, Reg Vardy Trucks of Reading, brown fluid from a mug is tossed out of the cab window and washes over my right Doctor Martin brown boot. A look of horror from the driver inside. His mate sitting alongside him wearing a motorway yellow fluorescent jacket and jackdaw teeth grins inanely.

'Sorry, son, didn't see you. You alright?'

11 March

Sunday afternoon. I am reliably informed that Princess Diana and the two princes, William and Harry, once called in for a Burger King on a Sunday afternoon. Does Charles drop in with Camilla occasionally?

Two white lads cruising the car park in a Ford Escort, no seat belts, music playing loudly, wearing white baseball caps, on the look-out for what?

Inside, three Americans buying flowers at the shop accost me. Tammy, Bill and Lulu are travelling around the UK on holiday. Tammy and Bill have been married for thirty-five years. Lulu is going through her third divorce.

I help them wrap up their flowers in the absence of any sales assistants – a gift for the local hotel manager who they say has treated them like family. I help them negotiate their way out and back into the car park.

Lulu squeezes herself into a 1970s white Mustang convertible, whilst Tammy and Bill go for the space of a middle range, imported Ford gas guzzler.

I log off and leave.

13 March

Early morning. I count six unmarked red vans in the car park as I pull in, one with a large sliding door open. A tall man in his late twenties, with long mangy hair, stands smoking a cigarette wearing a pair of Marks and Sparks carpet slippers which my grandfather used to wear at home when we didn't have visitors. He has a faraway look.

Men's 'sit downs' for a No. 2. It's completely full. I have to linger with some other men. We all avoid any hint of eye contact, just in case. Leaving my cubicle, a man waiting rushes past me as the toilet system is still flushing. I wash my hands pondering the fact that it seems to be the case that some men don't wash their hands. Alongside me, a man in his fifties, is shaving himself. His small rucksack, with British seaside names sewn on to it, sits by his feet. Is Paul Theroux off course?

Coming out, two Oasis-aged Mancunians desperately trying to look the part enter.

'Fucking give her a slap around the head.'

'Yeah.'

'Deserves it.'

Just catch sight of the mother with her son with the hearing aids leaving.

4: Shaggers

> This is one of the lessons of travel, that some of the strangest races dwell next door to you at home.
>
> Robert Louis Stevenson

The travellers are back – illicit relationship? – Mrs Philips and her sister – meet Willie at the Hair Bus Company – the grey heads are in – sing happy birthday to Sanjay – Spurs crew – Peg and Ron – Welcome Break lovers – the music here is good – Bank Holiday Monday – police tuck in to the burgers – a Buddhist – stories from the hotel – invasion of the small people – funeral party.

17 March

Saturday evening. The travellers are back in the car park. I count some fifteen caravans; their generators throbbing out a musical beat. The inside lights are falling out of the caravan windows. Not many cars and vans about. Are the men still working?

A police car, with its blue lights flashing, is leaving the car park. A group of coaches sits untidily in the coach park.

Inside a terrace, load of Tottenham and other clubs' football shirts are being worn.

'How did Tottenham get on?' a fourteen-year-old asks a younger boy.

'We won 3-0.'

'Fucking lucky,' comes the reply. An Arsenal supporter maybe?

The University of Kent sub-aqua team, five women, in their early twenties, dressed in tracksuits, pass through in a surreal procession.

SEVEN MAJOR TRAVEL AWARDS
Wallace Arnold
Wallace Arnold

SOUTH MIMMS SERVICES
Eat before Fuelling

18 March

The coach park is full with Wallace Arnold coaches. Once a year skiers mix with grey-haired 'oldies' off to savour the Lake District and other foreign parts.

I spy an illicit relationship. A woman in her forties, dressed like a schoolgirl in a short pleated skirt, glasses, a finely featured face, with expensive shoes and black tights, parks her Honda car. A man, about the same age, but with less hair, and wearing a multi-coloured sweater greets her from his blue Ford Mondeo. They hurry off together into the sanctity of a corner seat inside.

Mrs Philips and her sister sit at my table.

'We have used South Mimms Welcome Break Service Station many times over the years.'

'We travel with Wallace Arnold coaches, so four or five times a year we come here as the coaches change here.'

'We like the facilities, they're good.'

'There's plenty of room to sit down and it's pleasant with lots of windows which makes it light and airy.'

'You have to queue quite often at the counters.'

'The food is very expensive, don't you find?'

'The toilet seats are always clean though.'

My two eggs on two of toast take some time to be served. The staff behind the counter are being overrun. Roseanne, the middle-aged Scottish woman, has a set to with a supervisor. A poor unsuspecting man with two teas has queued at the wrong place. He gets the full force of the staff's stress levels. They ignore him completely.

In front of me a balding man in his late thirties, with a northern accent and goatee beard, is with his oversize wife and small daughter. They have the £6.99 traditional breakfasts with a side order of more eggs and toast. How much is it possible to eat at one sitting?

Orders are shouted through the serving hatch. Unknown persons inside shout back. A member of staff, a gormless-looking young lad, mouth open, pours hot water into a bowl of peas. He is oblivious to the increasing number of people who wait impatiently in the queue.

Three people, tall and willowy, all with long hair straight out of 1960s Carnaby Street are deciding what to purchase from the shop.

An unusual posh accent for this time of day is in range of my left ear.

'What shall I buy her, darling?'

'Flowers?'

'A box of chocolates?'

I am into my second cup of coffee. A Tory MP lookalike, complete with blue pinstripe suit, club tie, and distinguished grey hair, sits at a stool with a Virginia Bottomley double. Her blond hair may be turning grey, and her neat two-piece suit, with matching petite shoulder bag, does have a 'tea dancers' feel to it, but she looks the part. They survey the goings on around them with unbridled contempt!

'Darling, where do these people come from?'

21 March

A group of travellers is in; two, small, squat-size men, brothers, in their late twenties, both five feet tall, with small legs, with four children. Lively kids – one, a ten-year-old girl with carrot colour hair, is pushing a buggy with a younger child sleeping. They all look as if they are from the same family. They have been to the toilet for their morning wash. They sightsee. The children laughing and running around. They leave without buying anything, under the close scrutiny of one of the African security guards.

One of the table clearing staff is new! He is tall, sort of gangly, blond curly hair and dark glasses. Looks a cross between Buddy Holly and Chris Evans. Not in uniform, he patrols the tables with hands behind his back, gripping a blue J-cloth, like a 1950s 'Dixon of Dock Green' policeman doing his beat.

'Hello, hello, what have we 'ere? A couple of dirty cups and a tray? We'll soon sort you out.'

Out into the car park. The travellers have upped and gone! I tread carefully to my car to avoid the dog shit.

24 March

Decide to go walkabout. Venture up to the Hair Bus Company parked on the edge of the lorry park just up from the service station area. It's a rusting red and white single-decker bus with flat tyres. The neon 'Open' sign has long since died. Its replacement is in 12" high lettering and stuck on a side window like an oversize parking ticket.

I climb the steps of the M reg. National Bus. It's seen service in far more salubrious places. There's no driver to sell me a ticket. Instead, Willie, a short man with messy grey hair, wearing blue sport trousers with zips open up to the knees is cutting the hair of a Bruce Dern lookalike. His black tee shirt has a 'Save Willie' message on it. A pair of beach sandals hangs lazily on his feet as he moves around his customer with careful synchronised movements. Scissors chat away. Bruce is a trucker who's passing through. He's a regular customer here.

'I used to drive it around just to keep the engine going, but there's no point in moving it really because it's here all the time. It's become a bit of a landmark. I haven't painted it since, well, for years, a few repairs on the inside, but basically it just stays here.

'90% of my trade is truck drivers, the rest are people who work here, and a few locals who come in. Every day you get somebody new. There's so much passing trade here so it could be reps, it could be lorry drivers, whatever. The truckers, they're all nutters, the lot of them.'

A procession of middle-aged local men pop in for 'holiday' haircuts in between the truckers, as we talk. Willie knows them all by first name.

'What's it to be then?'

'No. 1.'

Paul sits in the chair. He wears a fluorescent jacket over his white lorry cab bleached skin. A gold cross dangles from his left ear. Dumpy driving legs swell out from beneath tight-fitting shorts. His short grubby off-white socks peep over unpolished black steel-toe capped boots.

Willie plays the stock market in between the snips. His mobile phone is constantly ringing. Calls are taken in his office behind the curtain at the back of the bus.

'I play a lot but I haven't got to the stage where I can say that I would be comfortable doing it for a living. The last couple of weeks I couldn't do a bloody thing wrong but then last week I got fixed for about six grand – that pissed me off. I made sixteen grand the other week but you could easily bloody lose that. To me it's just on the side, extra really.'

'He always says that. He's always moaning. He does all right; he just gets bored.' (Paul)

Another trucker dressed in blue overalls with 'Turners Bulk Liquids' on his back climbs aboard the bus in his slip-on sandals. His size matches the company logo. An extra stomach flops out from beneath a Fred Perry top.

'He's doing a survey of the sex lives of truckers,' Willie helpfully informs Joe. I smile.

'Well, I am not getting any of it.'

'No, he's from the Inland Revenue.'

'I am still not getting any of it.'

'You always get offered things. Years ago you could – not so much now as things have tightened up dramatically. But years ago if you said to me I would like a grand piano, you can bet your life that on Thursday somebody would come round with a grand piano on a truck. They come up with the most amazing things, sides of beef. But as you know wherever there's people, whenever there is anything in transit, there's always something going on; if they are not selling fags and beer, it's something else. All basically stolen goods and yet the police won't go anywhere near it and they know it goes on all the bloody time.

'They're had lorries out of here. See that concrete block out there, they had a lorry down there, a lorry parked across it; it was hemmed in on both sides and a concrete block in there and they still got it out.

'We used to have a load of hookers coming in at night, to service the wagons.'

'It still happens. A deaf and dumb one comes here regular.' (Joe)

'I used to open till nine o'clock at night when I had a girl working for me and you see the activities then because I had one come and bang on the door here offering. They just turn up. Where there's people there's money. They might work down the road and just come down here for a bit of extra cash. There was a guy, the one I saw a lot of, he used to drop them off in a minibus and then pick them up again. He was about six feet bloody four, covered in tattoos. I cut his hair.

'You get some lorry drivers, they would bore for Britain in the Olympics. They talk in a boring manner. Sometimes you wish you hadn't asked them a question. You ask them something and then they give you a blow-by-blow account of where they have been and where they are going. As if I'm interested. If they said I have just come back from Croatia and I have done something interesting then fine you can have a conversation and then if you ask them about the truck – and I am not interested in trucks – you get this whole nonsense about I drive a Scania, top of the range. It's sad in a way because they are locked up and they just pour it all out.

'It's not a life as we know it because most them belong on the road six days a week and get home for one night and back on the road again. Or if you are doing the long-distance stuff like Russia you are away for two or three months. They become a bit insular. They sit in their truck all day long and they get in here and there is an outpouring of grief or they don't say a bloody word. You can't hold down a relationship like that. You're on to a hiding to nothing if you are not at home. These guys are on their second or third marriages and they usually get taken to the cleaners and all they can be accused of is providing a home and they are the ones who get tanked.

'This kid came in and he was about in his twenties and he was slagging off women left, right and centre. Some woman had got hold of him and manipulated him and she had got like two or three kids from her previous marriage and while he was away she was screwing everything in sight and, of course, turned him over totally. He lost his house, lost the bloody lot and he hated women. Basically they need a pair of ears and it happens to me. They come in here and they off-load. Bit of an agony uncle!

'I never thought I would be here for ten years, it was just a laugh really. I'd done up the cottage and was messing about doing up some friend's house and you run out of what you want to do and I thought I would just go back to doing it to get some cash and pay the bills so to speak.

'I have been trying to get my old woman to go to Spain but she's always ummed and arred and said she didn't want to end up in that lifestyle, drinking earlier and earlier. I just want to buy a place out there and do something else. I get bored with this.

'Been next door to the service station? You must be joking.'

27 March

Coach park full of Wallace Arnold coaches. Holidaymakers with suitcases. Grannies and granddads walking around bumping into people. One elderly man walks into the disabled toilet thinking it's the way out. His 'minder' collects him. 'Couldn't read the sign,' I hear him whisper.

The rattle of a heavy machine gun coming from the Game Zone disturbs my early morning coffee and croissant. A man in a white puffed down jacket, baggy blue training trousers and white trainers, in his early thirties, and slightly balding, does his Terminator routine – it's 8 a.m. in the morning! What makes a man do this? And at this time of day?

*

Motorway Service Areas (MSAs) must not be used for purposes unconnected with the use of the motorway. The MSA must not become a destination in its own right, generating extra traffic on the motorway. Its purpose is to serve the incidental needs of people in the course of a motorway journey. (Department of Transport)

*

1 April

The Keith Vas lookalike is Duty Manager this morning.

People are tucking into the great English breakfast, a mixture of full Fat Boy breakfasts and bacon rolls.

The woman is in with her son with the hearing aids. She asks Sanjay where Chris is. The large female counter assistant who is serving the Fat Boy breakfasts brings over a handful of tokens for her son. He thanks her bashfully. Sanjay smiles at the woman as she orders her Cappuccino and tea, and bacon roll for her son.

I've finished my coffee and read Roy Hattersley's delightful critique of Blair's religious mouthings in the *Guardian*. There are only four of us sitting around – all men. Why is it so quiet?

Back in the car park. Two other VW cars have 'nestled' up alongside my car. Three white males stand around, leaning against a Kenning hired white van. A red soft-top Ford XRS Escort is the focus of their conversation.

2 April

Surprised by a man with golden hair, dressed in a white canvas suit, with brown suede desert boots, and a blue and white umbrella sticking out of a white duffel bag. Is he Number 6 from the Prisoner series?

Sanjay calls Chris back from the kitchen.

'What's this?'

'It's a birthday card, mate; Happy Birthday!'

'Thanks, mate.'

A spontaneous rendering of 'Happy birthday' breaks out from the staff and a few of us customers.

My table is alongside a group of Tottenham supporters who are going off to tomorrow's game against major rivals Arsenal; all men, grown-ups, in their mid to late thirties all of them. Dressed in smart, expensive casual gear, dark glasses. A couple of them are Jewish. They are all talking at once, just like a group of young boys off somewhere. Other men turn up. Baseball caps, dark casual gear and bald heads. *Daily Mail* readers. They fancy themselves as bit of a 'crew'.

Jason's getting them in.

'What do you want, Tell?'

'Bacon roll.'

'Get the teas in while you're there.'

Standing and sitting around me they are seemingly oblivious to everyone around listening in to their conversations.

'So where're we staying?'

'Not too bothered which hotel, long as the Old Bill keep away.'

'Where's the meet tomorrow?'

'Is he out now?'
'You're joking, he got three months didn't he?'
'You got to be looking at 400 geysers involved and he copped it.'
'Fucking geysers everywhere I tell yer.'
'He's coming out of Canvey.'
'What, going over the bridge?'
'Yea, it's only got one poxy way off the island.'
'Fucking hell!'

Mobile phones are keeping everyone in touch. It's a meet. Handshakes all round as men continually join up with the group. They are loud, aggressive and macho! People sitting on the other tables stare worryingly. A group of French families pick up the vibes and leave. Some are sitting, the others stand menacingly in the walkways. One, in half-cut dark glasses, stands behind me, mobile phone chattering away. I await a thump in the back of the head. Does he think I am an undercover police officer? My neck muscles tighten. I can hear his breathing and smell the stale lager.

'Yeh, so is Paul coming?'
Laughter.
'That's the dog's bollocks yeh?'
One man comes in, handshakes and respect all round.

New top of the range People Carriers are waiting for them in the car park. Some big geysers amongst them as they carry their overnight bags out to the waiting vehicles. One geyser plays 'Glory Glory Tottenham Hotspur' loudly on his car radio. Lots of Del Boy-type posturing.

10 April

Amongst the businessmen, the grey suits, the boxes of multi-folders, laptops and mobiles, another sighting of the man in all-white, and desert boots, and the blue and white umbrella in his white duffel bag. Number 6? He passes me, and I watch him walk out of the car park.

A young Coca-Cola sales team are having a meeting; all sit around a table – four women and a man. They wear black Coca-Cola fleeces with red piping uniform. An older woman is in charge, dressed in a two-piece light-grey suit. She seems nervous and perhaps new into a management role. She is the 'team leader';

it's her team. They all seem to be named Sarah or Luke. Expense forms are being carefully completed.

11 April

Peg and Ron are in their sixties. They live in South Mimms village overlooking the service station. They have been back in the village for two years now in the house that Peg was born in sixty years ago.

'We have been there twice, we never knew what it was like. We took our little grandson down there for a meal to Burger King. Been there a couple of times with him. He's eight nearly. We walked there – mind you, you take your life in your hands to walk through there. We walked down Greyhound Road, it only takes ten minutes but I had never been down there ever. This was twelve months ago. No, a lie, nine months.

'We knew it was a services because that's about one of the only on the motorway. We didn't even think about going there. It's only really for people travelling on the road.

'We went to The Crest Hotel, when it was called that, one or two times to have lunch but we had never been to the services. We didn't know there were shops down there. There's that shop that sells clothes, magazines and the size of it is huge. Never ever appreciated it was so big.

'Years ago in the village, we had four shops – Post Office, everything and there is nothing now. Nothing at all. No shops at all. Only the garage.

'I was very surprised at the size of it and the way it was laid out. We thought it was quite nice. If you are doing a journey, irrespective of whether you are on the A1 or M25 whatever, if it's so convenient to get to, I think it's a lovely place to come and stop and have a real good rest and break.

'It's better than those other motorway ones that have a bad name. It was very, very friendly and it was very, very clean. The staff are superb. We liked the sweet counter!

'I noticed there were as quite a lot of Army personnel came in there, and there were one or two police in there having their break.

'It's very modern and airy. They rebuilt it, didn't they? We went there after the fire so we don't know what the other place was like. It reminded us of the Millennium Dome. Not that I have ever been to the Millennium Dome! It quite surprised us, the size of it.

'If you were to ask 90% of the people who live in the village, they have never been or wouldn't have any interest in going. It's dangerous. They should make a proper walkway where you can go from the village and walk down to the Welcome place. You have got those lorries that come round the roundabout and fill up with diesel in the garage and there are hardly any pavements to walk that side and you take your life in your hands. It is not really very nice to walk to. If they had made that a bit safer I think you might get people walking down – well, kids would walk down.'

14 April

I am squashed between a table full of bikers dressed in black leather, and pensioners on a coach trip. Some of the latter are just getting acquainted.

'Where're you sitting on the bus?'

'Yes, at the front.'

Late middle-aged, overweight and all dressed the same, they sip their coffees and teas in unison eyeing the surrounding clientele as they do. Their coach driver sits away from them resplendent in his ruby-coloured jacket and matching uniform. His assistant, a woman, is eating a fruit salad trifle and drinks bottled water in between mouthfuls. He scoffs a full breakfast. She is dressed in the company tartan jacket. There is lively banter across the tables. One of the party, a grey-haired woman, bit like my aunt Vera, in a red full overcoat, sits alone quietly chewing her slice of cold toast, her red lips expressionless, eyes looking afar.

Across the way, two young lovers sit facing each other across a table. The table clearers try to earwig. They are off-duty Welcome Break workers just come off their shift. He talks quickly, she listens attentively. Both have jet-black hair. He has a moustache. She doesn't. She plays with her hair as he looks and talks. They are both dressed in dark cheap high street clothes. He is slightly balding and looks older than her. They leave together. Her worn blue jeans hang loose but tight across her bum as he walks a few paces behind her looking. They look a poor, sad couple.

18 April

Back to normal after Easter. Businessmen have returned. People passing through. Fat Boy breakfasts by the dozen.

Two church men sit alongside me discussing congregations.

'What do people expect?'

'What do they see in the church?'

It's a serious theological discussion and so early. I check my watch, it's 7.35 a.m. One of the men, the younger one, looks like the typical young male born-again Christian. Dark suit, yellow tie with black spots, white shirt, slightly balding with glasses. The other earnest fellow is a young William Haigh type, in sensible grey colours, and black well-polished shoes. Both lean on one elbow as they make their points.

A crowd of young Asian youths appears from the coach travellers' entrance. Noisy, shouting, vibrant, cracking jokes. They swarm into the building. People look uneasy, and seem to be physically keeping their heads down. It's two days after the Bradford race riot.

Paul McCartney's singing 'Yesterday' overhead. Little does he know it but he's competing with a group of parents and their little children eating large Burger King boxes of delights.

As I leave I am nodded out by a member of staff.

23 April

Two large truckers wearing grubby red boiler suits and trainers have sneaked in from the lorry park. They tuck into Fat Boy breakfast extras and read *Sun* newspapers. Both are oblivious to the 1980s Boy George classic 'Karma, Karma, Karma, Chameleon' which is playing.

Three men are facing inwards this morning whilst everyone else looks out to the car park. Haven't they noticed they're facing the wrong way? Should I tell them?

John Lennon's song, 'Just a Jealous Guy', follows. I hum the tune whilst four fat ladies, with west country accents, all 5'3" tall, waddle to the toilet miming the words. One of them has dyed blond hair with a see-through black lace-type blouse and black bra.

No male eyebrows are raised or heads turned. The discussion is about crisps, doughnuts and calories over cups of coffee.

A coachload of pensioners has disembarked, walking sticks and much needed hip replacements to the fore. They wear the Mark and Spencer's uniform with permed grey hair for the women, and slick flattened oiled hair for the men. Their coach driver follows at a safe distance, his beer barrel leading the way.

29 April

It's still raining cats and dogs outside.

A 'normal' family of four: Dad, with glasses, is in his late thirties – his dark hair is slightly balding; Mum is in black high-heeled shoes with silver stud trimmings, and a wine leather skirt – it's short, and topped with a skimpy one-piece with no back. Their two thirteen-year-olds sit quietly.

'Sugar?'

'Nah… Salt though.'

They have been joined by an older woman (his mother?), wearing a dress like my mother's curtains used to look like, flowery, and bright, but 'back room' curtains for the garden, not for the neighbours to see. Her hair is in a bow, and she wears glasses, with no make-up. Her feet are inside pink fur-lined carpet slippers!

Who chooses the music? The Beatles 'Ticket to Ride' is followed by a Cat Stevens classic, and then a Rod Stewart number. Hey – this is good!

It's very quiet now, only four of us in, all men, in the main seating area. A Jackson Five song follows next.

I can't help noticing a newcomer to the free concert, a big man, bit of a weight-lifter in his time, in green jogging bottoms and white trainers, with matching red socks. He has a Big breakfast with cow horns. I check my watch. Breakfast time passed some hours ago. The meal finished, he farts loudly and vanishes in the direction of the toilets. The opening bars of the Commodores hit, 'Please don't go', doesn't sound quite the same.

*

Then I saw a Great White Throne (Revelation 20: 11-15). For free copies write: Manny and Theresa Hooper, Revival, Missions and Evangelism, USA. (On the wall of the gents' toilet.)

*

4 May

It's Bank Holiday Monday. Religious leaflets have been dropped out of the sky by an aeroplane hired by an American evangelical church. I manage to avoid reading one.

All the retail outlets are open. The place is heaving. A couple of people dressed as clowns walk around. One is on stilts. Burger King boxes are in orbit. Lots of 'browsers', people wandering around looking. What is it with the British that we have to come to places like this as a Bank Holiday experience?

Some Hertfordshire Constabulary Special Force are in. Kitted out in all-in-one dark-blue suits they tuck into Fat Boy police canteen-size breakfasts all round. A mobile phone goes; possible action? It's a call from one of their wives.

'Hello... just got up?'

'Yeh.'

'My agent phoned. Do I want to do Kilroy? I said yes.'

'Just phoned up for a chat.'

His mates are making faces at him and go through a repertoire of indecent hand movements.

'Speak to you later, babe.'

Wandering back to my car, three dodgy-looking men are sitting in the front seat of an unmarked transit van, all staring intently at the entrance.

10 May

4.55 a.m. It's dark. The car park smells of Old Holborn tobacco and salty sea air. A lone bird flaps hopelessly across the near-empty car park.

Inside it's totally deserted. The shop is open but no staff to be seen! No newspapers, no croissants, and coffee served only at the main breakfast point. Two supervisors sit around on stools smoking. Standing on an orange Alpha Plant Hire extending platform three workmen are installing large TV screens in the seating area.

A ginger-haired woman, in her late twenties, wearing studenty glasses, purchases a coffee and hides away in the smoker s' corner. The in-house music pipes out its tunes to an empty arena. The night staff are strangers, no Chris or Sanjay. A cleaner-upper appears with a small child's satchel over his shoulder.

A lone truck driver breezes back to the toilet and out as 'How sweet it is to be loved by you', by the Four Tops (or was it the Supremes?), searches for some receptive ears. I feel lonely here.

13 May

Feel a bit disorientated and experience visual overload. The colour and noise of the 'cathedral' after the grey granite of Aberdeen where I was over the weekend hits the senses.

A bunch of middle-aged scousers have taken over three tables. They have that now familiar shop fitter's look. Tommy Cooper jokes fly around. One after the other they come – much laughter. The sound of crashing crockery from behind the table clearers' screened-off area brings forth howls of laughter and comments from them about 'fucking foreign shite'.

A William Morris type, young, is sitting by himself at a single table. Is he en route to Kelmscott? Or is he a Woodcraft Folk elder reconnoitring the service station for a possible future meeting point? The green cardigan, with his greenie-grey shirt, and light-brown combat trousers, and fourteen-inch beard are convincing.

A couple to my left. The man looks a bit like one of those English actors playing an off-duty Nazi in an early afternoon black and white BBC1 TV film, which nobody watches. His grey suit with a tangerine-coloured shirt, and pinky-red patterned tie go well with the old fashioned NHS glasses. He must be in his fifties. The woman is much younger than he is. His daughter? His body language says no! She wears a closely woven white jumper, fish-net style, not quite see-through, but close on. Her red-coloured hair is tied back, and she has a new age feel to her.

She leans towards him, hippy-style long skirt, in a brown and mustard pattern. They talk endlessly. Are they? Their hands briefly touch as they walk off to the shop. What's his name? Should I ask him?

20 May

There is a busy methodical hum about the place. People coming and going. Three armed Hertfordshire policemen are tucking into their Burger King meals, whilst seated on the uncomfortable stools which ring the seating area.

At the table across from me, is a familiar face – the 'Tourette's' man. He's dressed in a short-sleeve, blue Fred Perry casual type top, grey trousers, with brown easy shoes. His balding head bobs rhythmically as he drinks his coffee. He has a fixed stare about him.

'Georgio!' he shouts at the top of his voice.

I've been here before. It sounded like he was blowing into a conch shell as if his unearthly call is to someone amongst us as yet not identified. He continues to drink his coffee, placing it back into the saucer in three compulsive movements. There is no change in his facial expression. It's as if he had not shouted out.

People at nearby tables stare and look around. Two young men in suits giggle nervously.

24 May

Coachloads of over-fifties and school children compete for the space. A deadly mix. The former walk around lost, joining coffee queues at the wrong end and generally disrupting the smooth flow. The latter, mostly thirteen to fifteen years, noisily bump around displaying their 'gawkiness'. Bodies shuffle and crash into each other. Small boys congregate in the Game Zone area to play the machines. Yelps! Shouting flies across the hangar. Girls stand watching, giggling. Loud tuts can be heard from the 'Saga Louts'. What is it with these two age groups?

'Keith Vas' is serving behind the La Brioche Dorée counter, white apron over a large body. Well-spoken, educated English. I avoid any mention of possible misdemeanours.

*

There is food for all tastes – with high street brands such as KFC, Burger King or McDonalds. There are also traditional favourites from The Granary – or try the popular waitress service Red Hen restaurant.

*

29 May

A Buddhist is eating bran flakes, with no drink. His body is hidden under a blue fleece, walker's trousers and boots. A dark-blue jacket with a red slash sits in the seat next to him. I watch him take slow, deliberate spoonfuls of cereal, unhindered by what is going on around him. It's half-term so the place is littered with kids of all ages being slightly mad. He is not aware of them. A copy of Lama Surya Das's book, *Awakening the Buddha Within*, rests on the table.

The recently installed TV screens shine down on us like purveyors of some alien language, intent on re-educating us to the mindless subliminal sound of nothing! My free daily music concert of Seventies and Eighties favourites now seems long gone. Has anyone else noticed the change? Should I complain?

Three white men are with ten youngish Filipino women, all of whom have short dark hair, and smiley faces. They laugh collectively when prompted. Their eyes are fixed on one of the men, the middle-aged entrepreneur, with a dozen rings on his fingers, and an East Londoner second-hand car dealer's sales patter. His blond, beige-suited colleague, like something out of a whorehouse, sits there with a fixed smile. They are escorted by a Bruce Willis lookalike, sunglasses on his head, designer tee shirt, who tucks into his sausages. The small angelic faces watch both 'Bruce' shovelling down his breakfast and the rhythmic tones of 'Mr Big's' welcoming speech. Are all destined to be moved on as maids, housekeepers and assorted 'guest' workers? Or worse?

31 May

Azad serves me my coffee and croissant this morning. Lots of 'Yes, sir', 'Thank you, sir', 'Anything else, sir'. Feels a bit overdone. I find being called 'sir' very uncomfortable.

Chris fills up the filter coffee machine behind him, exchanging insults and greetings with staff who emerge from behind the door marked 'Private – Staff Only'. My croissant is colder than usual. Think about complaining.

The clanging of cutlery, plates, chatter wakes me from the sports pages of the *Daily Mail*. No *Guardian*s available for no reason.

I am surrounded by a group of Lancashire 'lads and lasses' of the over-sixties variety; grey perms, wigs, a bowls club baseball cap, and a motley collection of flat caps.

'Where's the coffee pot?'

'Don't get much do yer.'

'Was wondering if we would get served by that darkie.'

'Aye, they do.'

Talk of an attempted break-in to a car last night, in between doughnuts, eggs and bacon, and beans. Joe's wife gets his napkins for him. Ladies at one of the tables, men at the other. Blue blazers with gold buttons, club ties for the gents. Fred is in his orange shirt, and green and yellow tie under his blazer. He wipes his plate with his bread as his false teeth move up and down as they await the next mouthful.

1 June

The Days Inn Hotel is part of the service station complex. Peter is Acting Manager of the hotel. He's worked there for just over five years now. He left school at sixteen, and went to college to do a GNVQ in catering. We talk in one of the meeting rooms in the hotel. He lives at home with his parents.

'This is called Days Inn Hotel but really it's a lodge, just like a Travel Lodge but a bit better though.

'I started off as a normal receptionist and then Head Receptionist about two years now I guess. I generally supervise the reception staff, making sure they are doing everything right on the computer, making sure they are cashing up okay, any paperwork, stuff like that and just making sure they are getting on with everything they should be, really.

'I usually do 10 until 6, Monday to Friday, although sometimes I work the other shift at the weekends, fill in if someone can't work at all like a late shift, half two to half ten.

'We are quite cut off from the service area, main site. Most of the people over there have never even been over here; they don't even come in to have a look. I am actually surprised how few of them are interested in what goes on over here. We all have to go over there because we clock in and out over there, take our breaks over there, get all our food over there. We are all Welcome Break staff just employed in here, just like another unit.

'It's very hard to get people employed here as well. I think it's the positioning. A lot of the staff who work here come from – certainly in here – inner to North London and don't drive and it's very hard to get a bus here. The nearest train station is Potters Bar so you have then got to get a bus to South Mimms and then walk fifteen minutes all the way down from the truck stops. I get cabs because I don't drive. I am still living at home so I don't have to pay my rent but for other people who have to pay their rent it's very expensive to get here and what have you.

'Generally daytime staff last a lot longer than night staff. Night staff are very short-term. We have had so many people come and go through here. When I first started we had two regular guys who had been here a long time and then it all sort of went. A few people have come in, do a bit of training, decide they don't like it, or have an interview and don't turn up and are gone after two or three months. At the moment we have got staff. They have been here quite a while at the moment. If we ran out of night people now it would be very hard to get someone to replace them. Nobody wants night work. Most of the people on the main site just come in and do their shifts and go home. There isn't much connecting with us.

'During the week it's mainly business clients and they are fairly pleasant. Most of them have had a rough day on the road, stuck on the M25, and you can usually have a chat with them and they will be fairly pleasant. You get families mainly at the weekends and a few families during the week depending on school holidays.

'But at the weekend they mainly want to get in the room and get out, they don't want to stand around and have a chat, although we have a few regulars who will come in every week and we talk to them.

'Quite a lot of our custom is bosses and secretaries, or husbands and their mistresses. Certainly the weekends I would say 90% of our customers are what we call "shaggers"!

'Every hotel has got a different term for them, having been around a few. The tell-tale signs are fairly obvious. You can't always tell that they are not necessarily married but it's obvious what they are here for because that is why couples stay in hotels generally.

'They will usually always pay cash and if you ask for a credit card they won't give you that or they won't give you a proper address because they don't want anything sent there. It's a legal requirement for us to get an address so you have to have a go at them really and they never get anything sent anyway and they give in eventually. But it's quite hard to get a proper address. These are regular "shaggers". We have got quite a few of those. We have got a couple who come on a Monday and a Saturday, some that come once a week or more than once a week.

'We have had known drug dealers stay here in the hotel. Police helicopters and dogs surrounded the hotel. At one particular time I can remember armed police in reception – that sort of thing. I am not sure what the outcome of that was. I think the person got away actually. They were thinking they were still here.

'We have a lot of gypsies here, and they cause a lot of trouble. At the weekends they will try and get a room here. Sometimes they get through; not all our staff know the accents, a lot of them (staff) are African and they don't know the accents so they get in. We try and turn as many away as possible because we don't want them here. But they will get in sometimes and we will find in the morning that they have urinated all over the place. We have had them put their fists or kicked holes in the walls, and totally destroy the place, break tables.

'We have certainly had instances where it has come to checkout time in the morning and the gypsies haven't got out the room. So we have gone up there and told them they have to leave now, and they say no, we are staying, or we are going in a few hours, or what have you. We get the senior management over and eventually they come down and they will have a shouting match. We have had to call the police in the past but as soon as that happens they scarper pretty quick before they get caught.

'A young woman tried to commit suicide here. I found her. This was about a month or so ago, maybe two at the most. She checked in about half ten in the evening the night before. I wasn't here but the person who checked her in said she seemed a little off. So she was due to check out at eleven o'clock the following morning. I think one of our room managers came in, opened the door at eleven thinking she had gone and he'd seen that there was somebody still in bed so he just left it. We left it until about one and no one had surfaced so we phoned the room, banged on the door and our phones are very loud; you can't sleep through them. So we all came in here, me and him, and she was obviously still in bed and hadn't moved. Her leg was still hanging over the end of the bed.

'We were trying to shout and wake her up but she wasn't having any of it. I looked over on the bedside table and there were stacks of boxes of sleeping tablets. There was no visible evidence of any alcohol, just a bottle of Coke. So obviously we could see something was going on and I looked over on the bedside bench there and there was a letter, saying "Sorry you had to find me like that. Can you notify my parents…" – that sort of thing, so we called the ambulance and they came down and took her away.

'She sent a card a couple of weeks later saying "Thanks for doing that; I am okay; sorry you had to find me like that." She was from Leeds. Early twenties. White. Young girl. She didn't give any reasons for doing it. She left her bank statement on the side. She didn't say why she had done it but apparently she is fine now.

'We have had a few deaths in here as well, two that I am certain of. I think someone had a heart attack, an old gentleman, in one of the rooms and I would imagine the other one was that as well, no murders or anything, natural causes as far as we know.

'We had Sid Owen who plays Ricky Butcher in EastEnders

just the other week in the hotel itself. We had Ron Atkinson come in once on his way to one of the live football matches he was commentating. He came in to use the toilet and we don't have one in here for public use and he walked out pretty sharpish.

'We get people leaving things in rooms all the time. Mobile phone chargers is the best one. They always leave those plugged in the wall. Some sex objects have been left in the rooms. But just mainly normal stuff people would leave behind in a rush. Wash bags or that sort of thing, which are never claimed, of course. We have got a massive stereo upstairs somewhere which was left. People have also walked out without paying before.

'Eventually I plan to go to America so I don't want to be here forever. I was going to leave a couple of years ago but it has always been put off. I would hope that I wouldn't be here more than another year. I want to be somewhere else by then. I can transfer over to Days Inn America with any luck and that would be the easiest way to get out there and get a visa, but whether I stay in it for long – probably not. It's just a way of getting out there.'

*

Britain's drivers are being overcharged for poor motorway food and service.

*

4 June

Coachloads of elderly people, small women, 4'10" tall each of them, swarm over the place like a human tsunami.

'Where do we pay, Eileen?'

'I'll pay.'

'Hello young man.'

'Do I know you?'

Virtually every seat is filled. The shop is packed. Extra tills are opened.

Two elderly ladies on the table next to me sneak out their own sandwiches to eat.

The newly installed TV screens are quiet, their mindless message lost in a sea of elderly persons' chatter and crockery being nosily dropped.

A middle-aged man and his wife attempt to take two empty

seats at one of the tables. Embarrassed smiles from the table's elderly occupants. They are reserved seats for their friends from the coach. His wife squints through her NHS frames. They move on.

Two elderly sisters join me. Tea is served 'formally' by the eldest in her green coat. Danish pastries are cut and eaten correctly. Conversation is limited between the two of them. The younger sister has a French feel about her in her buttoned-up beige lightweight coat, tinted glasses, and small shoulder bag. Both have wedding rings on but feel as if their husbands are long since departed.

They talk to me like a long lost nephew.
'I like it here.'
'It's expensive but the tea is excellent.'
'We always stop here.'
Silence.
'Our coach driver has got something going with them.'
'Which coach are you on?'
I explain that I am not.

Later that evening…

The place is empty. Four young lads call in to collect a Burger King. The fifth member of their party sits outside in a clapped out yellow Ford Fiesta. Music is playing in the car at maximum volume.

I bump into a blond giant in the Game Zone. His red fleece, jeans, and big boots give him a hillbilly look. We play the machines together – strictly no talking.

A lone figure, an ageing skinhead, small build, 'tight' ears, eats a Fat boy breakfast.

The sliding doors bring in a collection of pre-midnight 'oddballs' from the outback. A couple of outbackers off the set of the film *Deliverance* walk through to the toilet. They don't reappear.

Outside in the midnight air the car park is quite full with cars. Some 'singles' are sitting in their cars expectantly, but where is everyone else?

5 June

There is a man standing alone on a beach to a backdrop of a blue sky and sea. It is silent; the new TV screens have 'frozen'. Peace reigns!

7 June

General Election Day. Do I care?

A family are passing through – they have been to Aunt Ethel's funeral – a grey-hair family of over-55s dressed in black. One of the men is wearing a dark blazer with his ex-RAF Squadron emblem on the breast pocket. The sober tone of the funeral is evaporating; some laughter fills the air. Of the five men and one woman, the joker sits at the end of the table. He tells stories of Aunt Ethel's younger years when she was a dancer.

The lone woman in the group, her sister, smiles like a counter assistant at Woolworths, as she gorges on a jam doughnut. A buck-toothed uncle peeps through his half-moon glasses.

'A good send-off.'
'Aye.'
'Didn't like the priest, too formal.'
'He was only doing his job.'
'Aunt Ethel would have enjoyed it.'

A coffee and tea refill is called for. Oblivious to the TV above their heads pumping out its retro music, they become reflective: a mixture of South London Battersea accents and the distinctive Thames Estuary English.

They stand to leave. Amidst the shaking of hands, slaps on the backs, hugs, and promises of 'We must all get together soon at Jack's'; a funeral service card is inadvertently left on the table. It is quickly collected by a table clearer and unceremoniously thrown into a black rubbish sack.

5: I was a Social Worker

Live the real jeopardy of circumstance.

William Least Heat Moon

New Labour are re-elected and there's no water in the gents' toilet – hangover – Adam used to visit here – is it cod? – Robert is off to university – join the smokers' club – Sanjeev confides in me – car park incident – find a wallet – bump into an Essex family – my car has a friend – 'close encounters' – bedtime story via mobile phone – pumping iron late at night – behind the 'Staff Only' door – Tiny likes working here.

11 June

New Labour has been re-elected.
'Sorry there is no water, sorry.'
'What?'
'You must be joking!'
A bad start to the day. Fire alarms are blasting out. There is no water in the gents'; the toilet attendant has difficulty making himself understood to the group of bewildered men, including myself, who wish to wash their hands.

One of the two entrance sliding doors is now shut. People coming in walk up to it and stare unbelievingly before walking through the other which is now permanently open. Two of the table cleaners watch and smile.

12 June

A group of elderly Americans lingers around the coach exit point, talking and laughing but perplexed by a group of young British schoolchildren on a school trip, who have stopped. The twelve- and thirteen-year-olds wearing jeans, sneakers and baseball caps look like American teenagers.

A man in a black casual shirt, with sunglasses on his head, is gesticulating wildly, his fingers pointing in all directions as if he's firing a gun. His blond female companion in her strapless turquoise blue top, gazes at him. The rat-a-tat-tat of the rounds being fired off is clearly audible as he boasts to the woman. A table clearer stands silently at a nearby table waiting to duck under in case a burst of fire comes his way!

*

BURGER KING: *Offering a quick service of tasty burgers including favourites such as the Whopper, The Big King, meal deals and more.*

*

17 June

The TV screens are turned off! A group of Sikhs in green and blue turbans, and women in long white flowing saris walk past the all-white heaving throng of coffee and tea guzzlers.

I have a hangover: a few too many Bangla beers with Jim last night.

The Ascot Races are on and some Londoners are all dressed in their 'best' en route. A blond woman, in one of the groups, in her late thirties, eats a double hamburger as she stands alongside men in their suits. Her backless high-heel shoes tip tap across the floor as they leave.

19 June

Someone tells Adam I'm writing a book. He phones me late one afternoon. We agree to meet up. A middle-aged man with a professional's voice, he speaks slowly.

'I was last there longer than ten years ago if I am honest. Used to go from St Albans after the pubs shut. It would have been between about midnight and two o'clock we used to go there because the pubs chuck out about half eleven and by the time we got there it would be about twelve o'clock. It was a Friday night usually.

'It was the only place we could think of that was actually open. There weren't too many clubs around in those days. It was post-divorce, post-separation, and I used to regularly go out and meet the lads in a similar position and we just didn't want to go home so we would go there. When we got there we actually found there were like-minded people there as well and you would sit and just start talking to people. You could just have a laugh and chat to people.

'A couple of us went regularly. It was quite a laugh. I wouldn't exactly say that you necessarily met the type of people you would want to have relationships with. So you go out there and meet some people and we got invited to parties and things like that. But nothing really ever came of it, like relationships developed, but in those days being ever hopeful was necessary!

'It was interesting. I suppose I was about forty but there were people there who were obviously eighteen or nineteen, and fifty or sixty. I suppose we were probably on the older side. I was very surprised when somebody told me about it. I said, you are joking, but I went along and it is true, people were using it just as a place to assemble. It was quite busy. They were from a surprisingly wide area. I distinctly remember meeting a couple of girls down from Hitchin, which is actually quite a long way.

'I thought it was good at the time; I thought it was quite smart. The service was efficient, there was a lot of cleaning going on, of course, at that time of the morning, but it was clean, it was okay – a darn sight better than some of the pubs we had just come from! It was life, it was something going on and I think when you are divorced if you go to a place on your own you feel strange, you just want people around you so it was an ideal venue for that. I haven't been back since, apart from to get petrol there once or twice. What's it like now?'

21 June

Bit like a gentlemen's club this morning; men alone at tables eating their breakfasts, the rustling of newspapers, middle-aged men with reading glasses. It's quiet!

Later...

A group of Taiwanese young people come in all talking at once! Two of the women are dressed in Simpson's jackets, plastic, red and green. They seem oblivious to people around them. A couple of the group, women, sit away from the main group. The table clearers stare intensely at them trying to comprehend their language, behaviour, and looks. An animated discussion takes place between two of them.

By the smell, a bald-headed baby has shit itself as Mum still prods it with some yellow-green substance on a spoon. The older baby with a bald head and Dad look on unimpressed.

Summer must be here again. The terrace doors are open. People are sitting at the tables outside. It won't last! A Vera Lynn lookalike in a pink halter top gives me the eye over a discarded Coke cartoon and Burger King wrappings. I gaze instead ahead to avoid eye contact.

Early evening...

A middle-aged couple negotiate their fish and chips dinner, bottles of fizzy mineral water by their sides. His distinguished grey hair and green stripe shirt matches his wife's fetching two-piece green outfit. Their knees are crossed in unison.

'Is it cod?'
Silence.
'Not sure.'
'It said "cod".'
'Here, try a piece of mine.'
'It's cod!'
'Are you sure?'
'No.'
'What is it then?'

I change seats and sit at the other side of the table. I get a side glance from a bonehead from afar. Feels like he is sussing me out as a potential fellow 'nutter'.

Two young members of staff wait patiently behind the counter for customers. The male, in a Welcome Break baseball cap, picks his nose. A supervisor springs out from behind the door marked 'Private – Staff Only'. A heated discussion follows.

Some 'parading' going on. Two Asian twenty-year-olds cruise backwards and forwards; goatee beards, white singlets, shorts and trainers.

A thirty-something-year-old white guy, tall with a red boozer's nose appears. His partner has on a white halter and false, suntanned legs. They're both Aston Villa supporters. Well, someone has to be.

Getting up to leave, a couple of male 'water buffaloes' catch my attention. Overweight with small thin legs, they carry off their Cokes and burgers to their cars. Two boy baby 'water buffaloes' follow them. The herd disappears through the sliding doors into the evening sun.

My car sits alone, moored up on the rubbish-stained tarmac.

26 June

The inability of the great British public to find their way around despite clearly designated signs such as arrows and notices is beginning to seriously worry me. I think Welcome Break need tour guides to both point customers in the direction on the toilets, the shop, the food outlets, the way out etc., and where necessary, physically take them.

'Mr Suburbia Man', wearing those horrible black slip-on shoes with the toggle which only golfers and over-aged East End 'wide' boys wear, especially with white socks and shorts, is walking about. He looks lost. I do nothing and pretend not to notice him.

His wife is flip-flopping about in a pair of dainty flower covered slip-ons that you see women wear living on Income Support in inner city estates or at Luton Airport. She has closely cut blond hair and glasses on a string! Her white tee shirt informs us all that she has just returned from a two-week package holiday to Ibiza. They stand just inside the entrance.

Later...

I talk to Robert in his raspberry-coloured Welcome Break uniform top. He's off to university next year.

'I can walk to work but I prefer to get a lift. It's about a quarter of an hour walk. I've worked here about two years in The Granary. They just put you anywhere. It's not like one place a day. It shouldn't be like that; it's only because we are short-staffed. The heat gets to you as well if you are behind the counters all the time.

'Some people are fine but you always get people who just don't understand. The old people are funny. When you see a coachload of senior citizens, they don't know where to join the queue, they don't know where the trays are, where the cups are.

They are like sheep. They follow each other. If one person is at the counter they all go over there and that one can be busy and that will be completely quiet. Safety in numbers I suppose. It would help if they had twenty minutes longer.

'It's surprising how similar people are and how they all say "Does that price include the cup?" or something, or "Do I get shares in the place?" They look at you expecting you to laugh. Some people are stupid. They don't listen to you. You say to them, "We only serve these meals now", "So you don't do roast chicken?", "No only these ones now", "So you have finished roast chicken?", "Yes", "There is no way I can get a roast chicken?" They don't take no for an answer.

'If you are by yourself serving twenty people someone will come up and say "There's no milk on the counter, can you go and fill it up?" and you think, no, I am by myself. I can either go and get the milk and have twenty people shouting at me when I get back. I call it the sheep effect because if one of them complains, they all complain.

'Some people complain about everything. We had one person about a month ago, he opened up his wallet and he must have had about £300 worth of vouchers from all different companies so what he does is go home and sends a letter and gets about £20 or £40 free vouchers and lives off them. He has a fat envelope full of them. If anyone complains they get vouchers automatically.

'Some people run off without paying. A guy who came in was totally high and he walked around the coffee stand for about five minutes trying to work out how the cups worked and he sat down without paying and Michael went over to ask him to pay and he couldn't work out what he was talking about, he was totally out of it. He was on something. We got security over.

'Was you here that night we had the riot police down a couple of months ago? Sort of street races or something. There were people coming from London in high spec cars and stuff and they were going to do some racing or something and forty riot police came down and closed off the whole of South Mimms and dispersed them slowly.

'We get some nutters in. We had a man stalking a member of staff. He had loads of money, quite a famous person and rich, and he eventually got one of his colleagues to ask her out. He was watching her for months, just coming in and seeing her behind the counter. You stand and look at people for hours and hours from here behind the counter.'

28 June

In the car park it's hot. 'Bob the Builder' dozes with his mouth open. His blue baseball cap and pencil RAF type moustache rise and fall with each breath. His ex-GPO red postal van is hand-painted a psychedelic yellow.

Walking in, a large dark brown rat runs over my right foot as I stumble along the path to the entrance. A hangover from last night's pub session in Hackney prevents me from reacting.

The lad with the hearing aids turns up with his brother. They both tuck into bacon and egg baguettes. No conversation.

I find myself gazing up at one of the TV screens at a *Blair Witch Project 2* trailer and return to earth to see a midget and her daughter take coffee: a middle-aged woman with dyed blond hair, a big head, and short arms, with her daughter who is taller by the three-inch heels on her boots. She's dressed the same as her mother. They disappear to the smoking area.

Lunchtime...

Caravans in great numbers in the car park. People giving water and food to their dogs.

I decide to sit in the smokers' area. Viewed from this end, the building is a confusion of colours, shapes and textures. The well-polished wooden floor shines brightly and the chrome and wicker chairs dance on it. It feels a bit like being in an old but post-modernist World War II aircraft hangar. Is that possible?

I feel different, almost superior, to the mass of people who sit in the non-smokers' area. Part of an exclusive yet dwindling club. Around me young businessmen are seriously puffing away. A group of staff sit at a table nearby, all women and all smoking. An older man, in an off-yellow shirt and dark glasses, sits in the corner observing everyone, including me. He puffs away at his cigarette and gives me a nod. I light up and join them.

4 July

Early morning. La Brioche Dorée not yet open. Get a pot of lukewarm coffee and a cold croissant from Corrine, who has no personality. She undercharges me. I don't query it feeling she might be too confused if I do.

The muted sounds of the awakening machines in the Game Zone rise above the sleepy gazes of the new arrivals. Even the Star Wars film music coming from one early morning punter playing a machine fails to change the mood.

6 July

Sanjeev has had a bad weekend. 'Problems at home,' he confides with me. He confuses my order. 'You usually have a coffee to take away?'

A family of French tourists lingers around the La Brioche Dorée outlet. An Inspector Clouseau voice impersonating Peter Sellers orders some doughnuts, three almond croissants and two pots of coffee. I am sure Kato is lingering somewhere.

Five young girls call in for an early morning Burger King. Two spotty faced teenagers behind the darkened counter explain they are closed. Astonished looks. All talking at once they noisily accept bacon sandwiches at The Granary. Two of the older ones in the first flower of womanhood gyrate in front of three men sitting at a nearby table, dressed in baseball caps, sweatshirts with hoods, and Adidas training trousers and trainers. They munch their way through 'fry-ups', mouths open, looking at the girls. I am reminded of three big hippopotamuses – it is not a pleasant sight. They are sitting there taking the piss out of fellow customers.

A middle-aged woman dressed like my Auntie Vi is turned away from Burger King; with her black shoulder bag she strides forcefully away.

The two young girls have joined the three baseball-capped men. One sits at the table whilst the other stands and fingers her midriff. Lots of talking and pointing. The long-haired one walks off to the shop, whilst the other confidently chats with them. I am feeling a bit uneasy about it all. Cigarettes are handed to the men. The girls are wearing very tight short shorts. They all leave together. I follow them out into the car park and go off-piste. Fearing the worst, I keep my distance but remain close enough. They all pile into a dark-blue Hyundai saloon.

The driver of the car, white baseball cap, sees me standing nearby jotting down the registration number. He mutters something to his two friends, gets out of the car and approaches me. It's not until he is up close that I notice the gaps between his teeth and that he is unshaven. To cap it all he wears a Stoke FC football shirt.

'What the fuck are you doing?' he breathes into my face. 'You police?'

He has moved within the 18-inch space at which it becomes threatening to the individual. I am caught out. I try an old Jedi Master mind trick to no avail.

Instead a previous life incarnation reasserts itself.

'I was a social worker and I am concerned about the two young girls who are with you,' I splutter.

I am chancing my luck here. He towers above me with an expression I've seen before. He pushes me in the chest with two dirty nail bitten hands.

'What's it got do with you. Fucking social worker, don't make me laugh.'

Can't quite place his accent. Still, that's the least of my worries. I mutter something about the safety of the girls, they being underage, my duty to do something about it. He laughs in my face. I feel an urge to knee him in the groin and put the boot in.

'Go on, fuck off!'

A couple of businessmen snoozing in their cars are now wide awake and look on bewildered.

A cigarette packet is tossed out of the window as the car roars off.

6 July

Haze from the sun-soaked car park tarmac lies like early morning mist in a country meadow.

Daniel, with a fresh crew cut and an undetectable accent, serves me. At £2.55 it's cheaper than the £2.80 I usually pay at the other Dorée outlet. The coffee is served in a cardboard cup. I decide to take the coffee and chocolate croissant and sit just inside the entrance.

Sitting in the 'enclosure' for customers, a black leather wallet stuffed full of credit cards is lying on the floor. What do I do? Hand it in or go on a spending spree? I carefully pick up the wallet and place it on the table. Somewhat paranoid that I am being set up like for a *Candid Camera*-type stunt, I slide the wallet under my paper considering my next move.

Two guys on the table next to me, one with a serious menacing Glaswegian accent. They see me writing. I smile unconvincingly. I leg it.

*

Excuse the Unusual Approach but you may well know of someone who would be interested in a Major 2nd income of £500 to £3000 per month Full or Part-time – From home. CALL 020 ------------- Normal BT rate – Call 24 hours. Please leave your name, address and telephone number and a complete information pack will be sent by post. (Business card left on the table.)

*

7 July

It's Sunday evening and there's plenty of hustle and bustle inside. The coffee queue curls like a snake. Only one person serving. I wait, but give up!

Three young women walk out of the gents' toilet, screaming and laughing.

A couple of bikers with matching helmets sit kissing on the table next to me. They share a Coke. A group of other male bikers, in their forties, with fat arses, wearing Frank Thomas boots, assorted Triumph bum bags, and ribbed black leather suits sit nearby watching the performance. Their FM helmets hang over their dark glasses. One wears a Jesus cross over his blue tee shirt. Helmets are taken off to reveal three balding heads. Burger King hamburgers all round. Their loosely strapped watches and wristbands flap as they eat.

A grey-haired man, in his fifties, is staring at me. With a milkshake in one hand, and chips in his fingers, he's rushing his meal and eyeballing every other male around him. I don't like the look of him.

A couple of sockless sandal people hop in; a man and woman, and a small child in tow. They ooze middle class, middle England. He wears shorts, which wouldn't look out of place in the 1950s. She's in check trousers with turn-ups. Their small genderless child is silent. They bring out their own food neatly wrapped in foil with the largest parcel for the man. A table clearer clocks them but decides not to intervene.

A 'Sunday father' is in with his six-year-old daughter. Their conversation is muted.

'How's the fries?' asks Dad.

No answer.

'Do you like the milkshake?'

No answer.

He clutches his coffee with both hands. She reads her 'Whopper Challenge Box' in which the meal came in. She has a pink hair-tie and looks like her dad. Her sleeveless top, green combat trousers, and serious trainers say he's not short of a bob or two for the maintenance!

Later...

I go walkabout and return to bump into an Essex family. There's five of them, including a bald baby in a high chair. The wife is blond, a long-haired Essex babe. 'All right.' The daughter and son both have skinhead haircuts, white vests, and big heads like Dad. They munch their hamburgers like a pack of wolves disembowelling a catch. Chips are shovelled into the boy's gob. Milkshakes are sucked dry by sunken cheeks. Father's Rolex watch is on display as he scoffs his food. Mum's open bag appears to hold some cups, saucers, and assorted items from the service station shop? Are they nicked? On reflection perhaps there is something to be said for Cesare Lombroso's theory of deviance?

Going for a pee I walk in behind Tommy Jinks and his son. He is thickset, his hair thinning, but still dark blond and in a ponytail just like when I used to play football with him on Hackney marshes when we were kids. His young teenage son wears a 'Duffer' sweatshirt with a hood.

'Cost thirty-five quid for that. Can you believe it?' Tommy tells me as he blocks the entrance to the toilet with his shoulders. He's a bookmaker now at a dog track in East London.

'Just passing, can't stand the place. What you doing here, Rog?'

My car has acquired a friend in the car park. A green Volkswagen Passat estate is parked neatly alongside mine. A middle-aged man and woman sit eating sandwiches out of silver foil in the front seats. A thermos flask sits on the dashboard.

They both smile at me as I climb into my car. I smile back.

12 July

Michael Portillo, the politician with his own website, has won the first ballot for the Conservative Party leadership with 49 votes. He denies he may be too 'radical' for the blue rinse brigade. More importantly I am politely informed that there are no chocolate croissants again!

A *Daily Telegraph* reader in brown brogues and a tweed jacket, with off-white trousers with regulation imperial 1" turn-ups sits by himself with his little pinkie sticking out as he holds his coffee cup. He reminds me of a gentlemen farmer I once travelled back with in a first class compartment on the afternoon Salisbury to Waterloo train. His black Freemason's ring shines under the artificial lights. With a military-style precision to his step he strides out to the car park.

Newcomers are now being confronted by three 'aliens' dressed as colour co-ordinated, Michelin-men sized Martians walking around for a photo opportunity. No one bats an eyelid as they walk about!

A tall, grey-suited, shaven head businessman, thirty something, sits drooling over his coffee with an older woman with blond dyed hair. She has tight trousers, and a black sleeveless top, and sling-back shoes with red painted toenails. Her full breasts pout at him and at anyone else passing their table. He fingers an empty milk carton as she teases him. She holds her cup with two hands and gazes over it at him. It's too early for all this surely?

My attention is distracted by a heavy Dublin accent. It's a frayed-at-the-edges Paddy, with a battered black, zipped executive bag, and arms like Popeye. Sitting across from him, a well-spoken Englishman. A re-financing package is being negotiated. Figures and sheets of paper are pushed to and fro across the table. The deal successfully transacted, small talk ensues and they go their separate ways. A mobile phone left on the table sings repeatedly for its owner until one of the table cleaners sweeps it up with a J-cloth and disappears through the 'Staff Only' doors with it.

As I leave, a fourteen-year-old man with an Adrian Mole face anxiously steps forward and asks me if I would assist him in a survey he is undertaking. He smiles nervously at me as I politely decline, mouthing an untruth that I have an urgent meeting to attend.

'No problem, have a good day,' he whistles through the gap in his teeth.

13 July

Two blond thirty-somethings giggle their way in through the doors. Sun-tanned, their bleached hair turns a few men's heads. One is in denim and trainers, with a Jill Dando look about her. The other, more vivacious, wears a light-green low cut top. She lounges back in her chair. A middle-aged man notices and fumbles with his laptop. Her pink toenails poke invitingly through her open-toed sandals. As she plays with her hair the men around seem afraid to look. A mobile phone appears, colourful like her white lace knickers which appear over the top of her jeans as she leans forward. A man grabs his *Financial Times* and exits to the gents'.

Later...

Travelling home from up the M25 I call in for a nightcap. A Pakistani Muslim family sits next to me. They don't buy any food or drinks but just sit talking. One of the women farts loudly. There is no reaction from the extended family. They go on talking.

It's almost 8 p.m. A lonely man sits and talks into his mobile phone. He's telling a bedtime story to his daughter.

Later...

It's 10.45 p.m. Empty. Chairs are neatly stacked on tables. Burger King closed. Both Game Zones are busy. An older man wearing a baseball cap joins the queue at The Granary. Only Lyn is serving; her hair is up in a bun. She takes ages to serve three

Nigerian women. The queues grow as Lyn serves them up a full fried breakfast. I check my watch. It's 10.59 p.m.!

A cheque card is produced. In front of me a Paul Simon-size man is becoming increasingly impatient. He cuts an unusual figure in his white sun hat, a pyjama-style jacket coat, and a two-day face stubble. He loses it; throws down his £2 coin and walks quickly off with his pot of coffee but no cup? Lyn apologises to me for the wait. I feel great warmth towards her. It can't be an easy job.

Sit next to four huge men with shoulders the size of American quarterbacks. 'Tough Talk' is emblazoned on their blue nylon jackets and tee shirts. They wear matching blue training trousers. The talk is of pumping iron. The older one has a North London Muswell Hill southside accent and tells the three about being 'Third in the world. That's official'. One of the mountains eyes up my puny frame. I fiddle with my notes and avoid eye contact.

The three Nigerian women are sitting nearby. They are talking in Yoruba, interspersed with some English. I cannot keep up with their discussion.

One of them has legs the size of a mature oak tree. She eats a bacon roll.

The guy smoking a roll-up cigarette with black liquorice paper has left. He was sitting on one of the stalls. Looked a bit out of place in his Adidas trainers with no socks. He sat there drinking a strawberry-flavoured soft drink.

As I leave I catch sight of a soldier, with two stripes, wearing an old, brown, very worn sleeveless leather coat. He reads a newsagent top-shelf magazine.

16 July

Had a bad classic migraine over the weekend with the flashing lights. Feel as if I am still somewhere else. Even the familiar sight of Chris handing me my croissant and coffee seems rather surreal.

I bury myself in my *Guardian* newspaper. The continuing Conservative Party election contest is making the headlines.

The middle-aged Asian employee patrolling the entrance and exit point in his dark blue serge waistcoat has an old British Rail stationmaster feel to him. Reminds me of trips to Blackhorse

Road railway station in Walthamstow with my grandmother who usually berated the platform staff about the dusty benches her young grandson was expected to sit on to watch the steam trains.

Aimlessly gazing into space, a dark-brown haired businesswoman walks towards me, hair in a bun with a black and purple ribbon, and a dark blue business suit. Her 2" high-heeled shoes give her that superior look. She occupies the 'round' table in front of me and is quickly joined by an older businessman who has that 'I am with a beautiful woman' look. Smirk. They quickly get down to sales figures, promotionals, and report backs. My mind begins to wander.

I catch sight of two women holding hands under the table. They see me watching them. I smile, they smile back.

20 July

It's raining and miserable as I pull off the motorway and into the car park. The second test match, the Ashes, is due to start at Lords this morning. Michael Atherton is captaining the England side in place of the injured Nasser Hussain.

Being away from here for a couple of days I feel the need to re-orientate myself slowly. I queue for my coffee behind a family of six footers with one of the sons wearing a number 5 football shirt with 'Luton Man' emblazoned on the back,

Why is it in a growing queue of early morning impatient businessmen that one of them produces a Visa card to pay for his egg and bacon roll and coffee? Is it the sensual, sexual pleasure of having his card wiped at that time of day? A high-tech blow job?

Get caught short, must be the coffee. It did taste a bit like the industrial cleaner Swafega. Could do with brushing my teeth.

Bump into a young boy in the toilet washing his feet in one of the sinks. He is by himself.

Afternoon...
I meet up with the General Manager of the service station. Shown through to his office in the inner sanctum behind the 'Staff Only' door. I am led through a maze of low ceiling corridors, which remind me of runways in a gerbil cage without the cotton wool. Photographs in cheap frames of happy smiling staff at team events periodically gaze at me from the walls.

The Manager has that quick 'jerky' profile of a man in a hurry. I am ushered into his office and come face to face with a wall of TV screens. Moving TV pictures of punters tucking into their meals stare back at me. He demonstrates how the cameras can zoom in on the smallest detail. Has he seen the grey-haired geezer covertly jotting down notes?

Did I know Princess Diana used to come in with her two sons, or that EastEnders has been filmed here? Elstree film studios are just up the road in Borehamwood where the soap is produced. Robbie Williams has sung in the car park. Do I care? Before the 'great fire' of 1998 people used to meet here before going off to 'raves'. Why has that stopped, were they caught in the fire?

The place had a big problem with drugs some years ago. Seems the police used to carry out black bin-liners full of drugs following raids. He tells me a story about the 'coach-spotter' who was caught by a group of soldiers jotting down number plates. Sad bastard! He was pinned down on the ground and had his camera smashed up. The police were involved. Appears he had some anorak hobby which led him to spend his time collecting number plates in coach parks.

'Security is tight,' I'm furtively informed.

The Canary Wharf IRA bomb was carried to its destination via the car park here. The vehicle involved was parked in the car park for two days. I was tempted to ask why the bastard who gave me a parking ticket for overstaying my two free hours didn't manage to get them as well.

'We have Portuguese-speaking staff clearing the tables, and the security guards are French-speaking Africans from various countries.' I nod with an impressed look.

Coffee appears. I make a joke about being a connoisseur of the coffee here.

'All tastes the same to me,' comes the reply.

I am given the guided tour of 'backstage' and eventually I am led out through the twisting corridors to the main area. The noise, colours and space of the outside world startle my senses. It's as though we have come out of the player's tunnel at Wembley. I move off into the crowd. A group of face-painted fatigue-dressed soldiers come at me. I skilfully evade them and move upfield. I am back in action and heading for the exit to be back on camera.

23 July

I am sure I can feel the gravitational pull of the north sitting here. Is this a hinterland where the service station acts as a fort on the most southerly section of Hadrian's Wall now called the M25?

Again there are no chocolate croissants! I settle for a boring plain one. Sanjeev shares my disappointment as he serves up the croissant. He looks embarrassed as its curled ends stick up like a Viking's helmet.

A long queue of men at The Granary for breakfasts. My attention is drawn to a newcomer to the queue in a black leather jacket, boots and jeans. He has lots of bits hanging off his belt, hair and his beard. He looks incongruous in the queue, an interloper amongst the 'normals'. He settles for toast. Much movement in pockets looking for money. He sits down facing me. His black headband gives him that 'missed the last ferry back from the 1969 Isle of Wight rock festival' look. I find myself unable to avoid watching him eat noisily with his mouth open and examine his nicotine-stained teeth and old fillings.

30 July

Tiny has black sticky-up hair. The last remaining tooth in his mouth hangs like a white stalactite. He's forty-two and lives out in the nearly village of Welham Green. He has worked here for fifteen years.

'Came here in December '87. I do maintenance. I look after the Days Inn Hotel, the forecourt, the truck stop, and the main service area. I used to do plumbing and heating all round the area. I don't really talk to the public a lot but I do now and then if they lose their cars.

'It only takes me five minutes to get here. There's no transport from Welham Green to here. I always drive in. I do flexi. I always do eight to half four and then weekends I do seven to half three.

'When I used to go round the car park, I used to do the sweeping out there on a big sweeper and used to find things lying around, like bags and that. I've found a few cameras now and then.

'They nick cars and dump them here. They drive all the way round the motorway just to get here to dump them because it's so easy. People leave their cars here. We always write to them. We are the ones who have to send off to Swansea and then they write them letters. Some of them do tell you, write back saying, no, keep the cars and send the log-books and that. We are not allowed to get rid of them until we have had authorisation from them and then we just get a scrap-bloke to come in and see if they are worth anything. Mind you, if they are there too long they will get stripped down; there are about three or four out there now. You've probably seen the ones that ain't got no wheels on.

'It's gone pretty quiet recently. Last week we were busy because I empty the car park meters and so you know roughly if you have been busier one week than others and you can tell if it has gone down a bit.

'All the businessmen have all gone on holidays. At this time of the year it will be pretty quiet on weekdays but it's always busy at weekends. Today it's quiet but there again you have got a pile-up on the motorway. So when that happens it's a little bit quieter because of that; no one can get round it.

'They had a break-in Saturday night in the Game Zone. They broke into about seven or eight machines. They are sitting there and as he is playing it he's breaking in. The others are watching, looking around thinking there's no cameras. They broke one machine, then another and worked their way round. They're all on camera, got their faces lovely. But it's just amazing. I suppose people hear them or see them and don't say anything. Two coloured lads and the other two, whitish or more Asian they reckon.

'I like working here.'

6: Breakfast at the Red Hen

> I am most entertained by those actions which give me a light into the nature of man.
>
> *Daniel Defoe*

I return – Brian Moore has died – John Betjeman is en route to Uffington – do we need a 'bum Tsar'? – an embarrassing moment – Van Gogh's workroom has appeared – bogeys whilst you eat – full English breakfasts – Post September 11 – coach drivers – Rev. Dave – Mr and Mrs Sharing – John, ex-Spitfire pilot – Manchester United everywhere – who's baby is this? – text message from Jim – Burger King is closed.

3 September

Sunday afternoon. I'm back. Been working and travelling in foreign parts. It's like I've never been away. The gents' toilet is as busy as ever. Bit of a rugby scrum to get in. Grunts and groans coming from the cubicles and the usual quota of men and boys not washing their hands after fumbling with their manhood. The place looks as if it's been steam cleaned.

Get short-changed by Sandra (a new face?). She blushes and has to get Andrea, the supervisor, to open the cash till to pay me. Coffee tastes familiar. Catch the whiff of a Chicken Royal Cheese Meal coming from Burger King.

A middle-aged woman, with grey hair in a fringe, is pushed in by her husband. People around stare as she eats a chocolate cake. Yes, I know she's disabled but she can eat the same as you and me. I think it but don't say it.

First day back and the tinny drumming from the hanging video screens is getting to me already. A black and white girl band is doing its best to wind me up. An older man with a walrus moustache is nodding at his wife. Their feet dance in tandem under the table to the music.

Five deep at Burger King, mostly under 25s. The post-match euphoria of England whacking Germany 5–1 last night seems to be reflected in the consumption of double burgers, large fries, and Whopper Coca-Colas.

I am in danger of becoming transfixed by the Welcome Break news beaming my way. Brian Moore, the football commentator, has died. I remember Sunday afternoon lunch after playing football for my local team, straight in the pub for a handful of pints and crisps before they closed at two o'clock, then home to my mum's roast beef and two veg, watching the Big Match on ITV with Brian Moore commentating with a belly full of beer and being knackered after the match – fond memories. Thanks for the memory, Brian. Bet you never thought your face would be beamed across a service station area?

I am not the only man wallowing in nostalgia. Alongside me a man in his fifties is tucking into a traditional Sunday dinner complete with Yorkshire pudding, two slices of bread and butter and a cup of tea.

The Welcome Break news is now telling me how many 'hits' a Miss Piggy website has had.

I begin to feel disorientated. I stare at the disabled woman. She has a voice like a late night burglar alarm, shrill and insistent.

6 September

A little boy with a pudding basin fringe is pulling at his mother's breasts. Her husband sits impassively, dressed in an English bee-coloured sweater with matching beard. He's a dead ringer for one of the three musketeers. They drink their coffee with delicate movements as if they were somewhere more up-market. Conversation is limited to shrugs of the shoulders and facial movements. The little boy looks as if he is demanding to be breastfed. A young hyperactive twenty-year-old, with diamond socks and designer glasses, drinks from two bottles, one water, the other a can of Scottish Irn Bru. He gesticulates widely as he speaks on his mobile phone. His feet tap continuously like a flamenco dancer in synchronised movements between his limbs and his voice. I watch, spellbound.

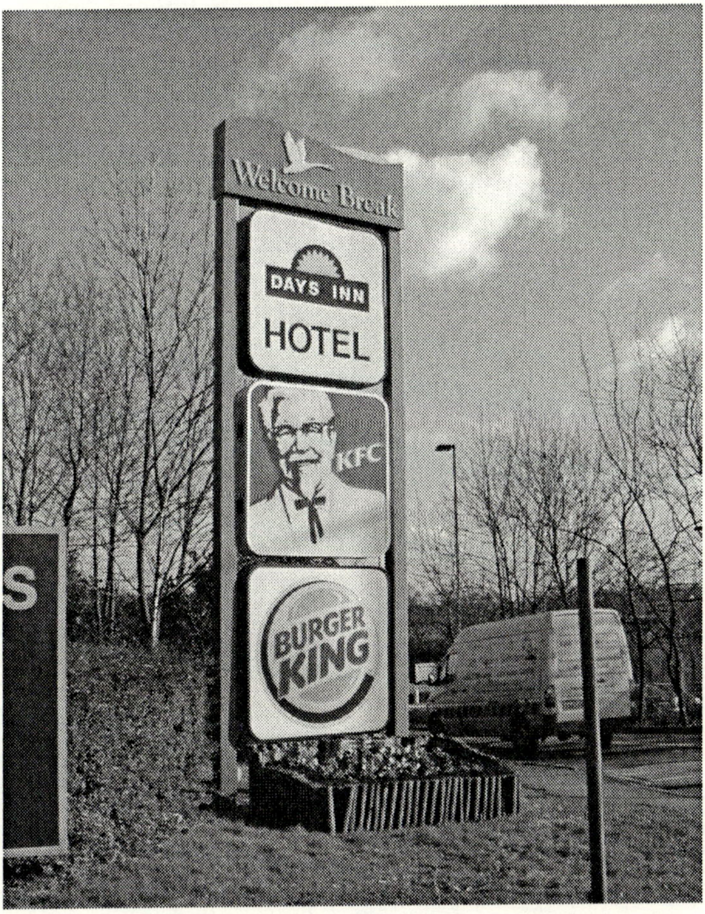
Welcome Break
DAYS INN
HOTEL
KFC
BURGER KING
S

An old man, John Betjeman face, with a flat checked cap is munching slowly on a Burger King meal. He is eyeing up people as they pass before him. He chews relentlessly, seemingly never swallowing. Have you noticed how men over the age of sixty still carry white handkerchiefs in their breast jacket pockets? He wipes his nose, polishes his hands and cleans his glasses in a rotational movement. He can't quite decide whether to drink straight from the carton, which is as big as his head, or suck at the straw. His hand periodically covers his mouth as he conceals a touch of flatulence. His baggy trousers and NHS walking stick gives him a Charlie Chaplin-like walk as he moves off towards the exit accompanied by his daughter, who has appeared from nowhere to escort him out. Home to Uffington for tea?

I move to the near-deserted smoker's enclave. A whiff of cigarette smoke mixed with an assertive pipe smoker means I have to breathe in short bursts to avoid inhaling. Why is it smokers, particularly men, hold their cigarettes either two handed with elbows on the table or singularly with one hand that moves the cigarette rhythmically up and down like having sex? Funny how being a non-smoker now, I enjoy the occasional passive smoking.

The place suddenly becomes bereft of people. Wide gaps appear. I begin to feel like a stranded whale on a beach. A man with pens in his shirt breast pocket eyes me intensely from across the tables, his coffee cup hiding his thoughts. He sports a tie with a military insignia. I notice his highly polished black shoes standing to attention under the table. He's not wearing any socks!

Later...

Two men in glasses are playing with toy dolls. Yes, dolls with spindly legs and arms. Are they sales reps or a couple of nonces? A little boy joins them. They sip orange and discuss the toys in heavy Dutch accents. They seem to have bags of them in different colours and shapes.

A 'more cheese, please' shout from the direction of Burger King disturbs my interest in the doll men which is okay as I am feeling a bit uncomfortable seeing two middle-aged males finger girlie dolls, particularly as one of them is wearing side-zip black Cuban heel boots. Who wears them in this day and age?

I get squeezed buying a bar of chocolate from the shop by people, both men and women, with big, sticky-out arses. Why is that? Whether in jeans, dresses, trousers, skirts or shorts, it's as if people are carrying around either their entire life's savings or an emergency food supply in their lower backs and beyond. Take this fellow with an arse the size of a prize-winning bloated overweight melon. He slowly walks in with a woman who would get lost in his rear cheeks. Wearing faded denim shorts doesn't help. Or how about the blond woman with a heavy black guy in dark glasses, who has a pair of multi-sided elasticated jogging bottoms the size of an ancient Redwood tree. Who was it who said we need the 'Naked Chef' Jamie Oliver as Britain's 'food tsar'? Perhaps we need a bum tsar instead.

*

Red Hen. A new waitress service restaurant, with a wide range of fresh, country style cooking. A great place for a full English Breakfast any time of the day. Refuel your appetite with a welcoming Red Hen meal.

*

7 September

Bump into an old football mate of mine. Haven't seen him for donkey's years. Suited and booted, his hair is tinted black to cover up his greying over. The beard is neatly trimmed. There he is sitting face to face with a much younger woman, pretty face, not too heavily made up, sort of face my mother would have liked, dressed in a two-piece business suit and matching white blouse. He looks up startled.

'Oh, it's a business meeting, best place to meet, you know, most convenient. What you doing here?'

'Just passing through, came in for a cup of coffee.'

Embarrassing silence. He doesn't introduce the woman; she is looking at me, smiling.

'Roger's an old friend of mine, we go back years.'

Still no introductions. I feel the need to escape.

'Well, I must be going, things to do.'

'Yes, of course.'

'Keep in touch.'

'Good seeing you.'

I pass out of the sliding doors into a cloud-soaked sky. It's cold but I am feeling hot and my cheeks feel flushed, I am blushing. Start thinking about Carlo Collodi's marionette Pinocchio as I walk to my car.

8 September

A new money-spinner has been introduced to the walk-through area. It's peculiarly named 'Van Gogh's Workroom' although any similarity between the Dutch master Vincent van Gogh and his studio at Arles, in the south of France where he produced arguably his greatest pictures, and this photo-booth sized construction is zero. Seems you can create a portrait or caricature of yourself and get the finished work of 'art' back in three minutes. There are no takers!

See my first Manchester United football shirt of the day. A number 4 Paul Scholes on a blond-haired man in builders' boots and blue jogging bottoms in with a mate. His mobile phone is moved from ear to ear like a trowel over quick-setting plaster. He has that knob first swagger and swinging shoulders. Must be a Londoner!

'Alright, son.'

'He's 'aving a laugh.'

The Welcome Break news is having an adverse effect on me yet again! A mixture of film previews, adverts for Chinese-style chicken legs, Dr Pepper's drink, and McVities chocolate Hob Nobs biscuits has me reaching for my Kalashnikov AK47.

Paul Scholes is eating a 12-inch roll filled with assorted meats. His jaws work through it like a car crusher in a backstreet East London metal scrapyard. One of the table cleaners breaks the 'number one' customer house rule: 'Do not clear the table until the customer has left.' He has been sitting and eating for so long that to avoid his table becoming overcrowded with food empties they helpfully make space for his next course. His newspaper acts as a napkin between bites. His mate looks on.

Why is it that some men pick their noses when eating? Is it something we learn as children? 'Now eat it all up, including the greens, and don't forget that nice bogey up your left nostril, there's a good boy!'

'Yes, Mum.'

It's fascinating. There he is sitting across from another man, mobile phone at sleep on the table, eating a sandwich, talking, and picking his nose. The guy facing him makes no visible recognition of this. If I'm tucking into an over-priced cardboard sandwich made with yesterday's bread and lifeless ham and cheese, the last thing I need is for a bogey to emerge from someone's nose on the end of a finger and disappear on their mouth along with a mix of bread, cheese and ham. Decide to give lunch a miss today.

Two people, a man and a woman in their early twenties, have entered into the Van Gogh's Workroom. Both are giggling. The woman, her tits pointed outwards and upwards, in high-heeled boots, carries a packet of salt and vinegar crisps. They emerge three minutes later. He has a new Levi's denim uniform and German blond hair. She munches crisps as they wait, and wipes her crispy fingers on her jeans. The prints eventually appear to much laughter. I just hope Van Gogh can't see this going on.

12 September

Early morning coffee.

Six 'Bob the builders' tuck into their full English breakfasts – mixture of baseball caps and closely cropped bald heads. Toast and jam followed by loud belches.

It's the morning after the terrorist attack on New York. Fifty floors of one of the World Trade Towers destroyed were rented by the international banker, Morgan Stanley. Emerging from the gents' a tall black salesman tries to sell me a 2.9% APR Morgan Stanley credit card. Life carries on. Surreal!

There seems to be a quiet calmness this morning, people taking their time, not rushing, floating in space, quiet conversations, pages of newspapers being turned deliberately.

A group of young Japanese tourists eating sandwiches and drinking Coke have newspapers spread out on the tables reading about yesterday's attack on the USA. They look nervously around them, feeling strangers in a foreign country.

Susan serves me my coffee and croissant aided by a new assistant without a name tag. Problems with curdled cream in my coffee. Des comes to the rescue. He fussily fiddles amongst the creams and milk cartons, separating out the 'bad' from the 'good'.

A few of us punters stand around joking as we wait for some 'good' cartons to be hand-picked for us.

More people than usual are glancing up at the hanging TV screens trying to wring out the barest of news about the situation in the USA from the Welcome Break news television channel. Difficult to do when the weekly film review in conjunction with Blockbusters Video takes priority.

14 September

Car park. Eight Club Cantabrica coaches are collecting people for sunshine holidays; kids happily out of school, giggling sixty-year-olds in shorts. Brutus, Popeye's old rival, leans against one of the coaches eyeing up potential Olive Oyls. The cigarette smokers stand to one side in a huddle.

A fifty-something-year-old woman with long brown hair, skimpy top and wriggly arse turns a few heads.

A family of skinheads, including Mum and Dad, carry pillows on to coach Number 2; little boys picking their noses and eating crisps, and men with Chelsea football shirts over big bellies.

Inside, three middle-aged coach drivers in blue blazers and ill-fitting polyester white shirts with wet armpits, and matching company striped ties, take a lunch break. Their comfy black slip-on shoes with matching white socks are obligatory.

15 September

The Reverend David is the vicar of South Mimms Village Church, an ex-Home Office civil servant with wonderfully tuned Oxbridge vowels. The vicarage is located in a small close of newly built houses some walk from the church. His study is small, warm and cosy, and reminds me of a hedgehog's nest. David is an historian. One wall of his study is a bookshelf. He wears handmade expensive brown brogues, long black socks and a Marks and Sparks jumper over a checked shirt. His off-white casual trousers blend in with the carpet. Coffee is served in delicate china cups.

'There are a few people who feel the motorway doesn't exist. Welcome Break is seen as the alien presence in the community.

'I have been to the service station but not very much. I used to go down there with my daughter who had a summer job there, to pick her up and take her down. We lived in Potters Bar before we came here, so she needed lifts.

'I cycle down there sometimes if I want a newspaper, and you can't get stamps here in the village, then you have to go down there for stamps if you run out. Or I walk down there, but you have to pick your way between the ruts the lorries have made, and getting across from this side on to the main service station there aren't actually any footpaths down there from the village.

'Occasionally refugees are dropped off. There have been people dumped there and wandered up here to the village looking for help. We had our local paper carry quite a spread on a family that turned up in Blanche Lane having been dropped off at South Mimms. I must say I was rather pleased with the way the village responded to that. The secretary of the village hall opened up and provided the means for them, people rushed round with cups of tea, and they spent most of the day waiting for the police to come and collect them and take them to a shelter.

'I have to say when I got your call to say could we talk about the service station it touched on a raw nerve because I am conscious that I devote my energies to these two villages but there's a whole of human life there upon our doorstep with which I have virtually no contact.

'I guess one could say I will become the unofficial chaplain to the service area and spend my time around there wearing a collar and perhaps being useful. There's an awful lot happening there. I've always had the sense that there's a whole world down there.

'Months go by, one year turns into two and turns into three, and I think still I have not made any moves into the service area. I think I will visit there more often.'

18 September

Long queue in the shop. Three white guys are buying a cut-price Verve CD. Money and credit cards change hands between them before the purchase is made.

I get elbowed out of the way waiting for my coffee. He's wearing a yellowish-brown khaki sun hat, a grubby Calvin Klein tee shirt and jeans smeared in white paint, supported by a pair of well-worn Nike trainers. His two pals are in hooded sweatshirts and faded green combat trousers. They collect their coffees and drink them loudly. Lots of laughter.

'Fucking tosser!'

They're having a laugh at people sitting away from them. Tables nearby remain deserted. The one in the hat sits on a stall to have a smoke. Plenty of ribbing about shagging a woman. One of them is into what Desmond Morris calls 'spreading their legs wide in a primate crotch display' as he pours his coffee. His mobile goes off.

'Yeah.'

'Leave it out!'

A coffee pot lid comes off, coffee everywhere.

'Fucking hell!'

'No.'

'You must be joking!'

'What?'

'Yeah, yeah.' Hand signals to his mates indicating the person is a wanker.

'Sure, see you later.'

I feel the overwhelming desire to flatten him.

24 September

Went cycling yesterday with some mates through the Essex countryside. Ended up having a pub lunch washed down with some locally brewed bitter. Paying for it this morning as I hobble in from the car park which is smelling of Monday morning and burning plastic.

I go up-market and take a coffee and croissant at the La Brioche Dorée outlet. Two middle-aged men sit alongside me at the next table. A 'Mr Marks' and a 'Mr Spencer'. A small plastic green box, the type that four-year-olds use at nursery school, sits on the table between them. It's one of those card index systems you can pick up at shops like WH Smiths.

Two Red Beret soldiers complete in battle fatigues walk in, both with that 'no brains but dangerous' look about them. They both miss the 'Miss Welcome Break woman of the month' who walks in behind them. She purposefully strides around and looking for Mr 'Lucky Bastard'.

A lull in the proceedings so I check my Saturday lottery ticket. 'It could be You!' has now reverted to 'Never in Your Life Mug!'

A blond woman in a two-piece grey trouser suit pulls in a portable 'Can I interest you in a new credit card, sir?' stand. It's a large suitcase on wheels which would have even had difficulty passing as hand luggage on a BA Heathrow – Dakha flight. She is also carrying a rucksack which seems to hold provisions for a month's duty by the toilet block where presumably she will be standing for the rest of the day. The rest of us stare as she struggles through. No help is offered.

I am having some difficulty with my second cup of coffee. It's tasteless and I am beginning to resent the £1.35 I paid for it. Should have gone to my usual outlet with Chris and Sanjeev.

The 'Enforcer' tells Rose and her colleague to clear the tables. It's a fair point as I am having some difficulty negotiating around the discarded cups, wrappers and empty milk cartons on my table.

Pauline is the 'Enforcer'. A middle-aged woman, with a serious attitude, who always dresses in a blue uniform trouser suit and waistcoat, and a 'boyish' haircut. She means business.

A couple are sharing everything at a table behind me; their spoon, the same page of their newspaper, and their coffee. They strike up a conversation with people on the table alongside me.

'What's her name?' Mr Sharing asks.

'Hannah.'

'Are you the father?' Mrs Sharing has joined in.

'Well, you never can be sure these days.'

A sharing smile, the thirty-something couple smile back. Baby Hannah stares and gurgles at Mrs 'Sharing'.

'My husband Brian would like to take you home with him, wouldn't you, Brian?'

Brian smiles sharingly as the baby grabs Mum's breast. Brian is a dead ringer for John Major, that smiley 'the lights are on but there's nobody at home' look, remember? Mum disappears with baby Hannah to the ladies'.

Mr Sharing now has his arm around Mrs Sharing. Their two chairs are close together, he turns the newspaper page over for both of them, she now puts her arm around his shoulder, they both laugh together. A La Brioche Dorée napkin is shared between them. Heads lean against each other's. They are both old enough to be my mum and dad! A pen is produced for the crossword. They share a single pair of glasses.

Tom Hanks is beaming down at me from the overhead TV screen, it's a trailer from his film *Castaway*. I wish I was with him. I have a great desire to separate the two of them forcefully!

Brian turns and smiles at me as he stands up to leave. Mrs Sharing follows. I am sure I hear them discussing 'Was he the father?' as they leave.

I exit into a darkening blue sky with a half moon peeking down at me through the artificial halogen lights of the car park.

'Is there a toilet in there, mate?'

'Is it busy in there?'

Two questions.

'Yes,' I reply, avoiding eye-contact with a man with a serial killer look about him. I think about dialling 999 but leave it.

25 September

Relieved to discover there were no shootings here last night.

The aroma of brown sauce over bacon, eggs and toast is overpowering. I move closer to the table it's coming from. My croissant and coffee pale into insignificance.

George Mowbrot has written a telling piece in today's *Guardian* newspaper which supports the anti-war views of Naomi Klein. Sometimes, since the days after September 11, seeing people going about their daily business has an unreal ring to it. No, there's no biological or chemical gas masks hanging over people's shoulders as they drink their coffees or like the guy tucking into his bacon and eggs with sauce, but somehow everything has changed – have you noticed it?

People seem to have a resigned, 'let's see what happens' look about them or is it simply we are all at the mercy of forces beyond our control? Are we all waiting to become victims? It feels serious. But life goes on. Tony Adams is injured again and out of the forthcoming Arsenal Champions League match against Panathinaikos in Greece. Now that's serious!

I drink my coffee up, nod to the bacon and eggs man who surprisingly nods back, and go out into the morning fog or whatever it is out there.

26 September

Bit of a kerfuffle at the counter this morning. Chris and Sanjay are overwhelmed by a demanding family of four. A mum and dad are taking their two teenage children off to university.

Sit next to a young white guy in his early twenties who has succeeded in emptying most of his cornflakes and milk on to his tray rather than the small dish. He sits with a group of landscape gardeners wearing green tee shirts and brown boots. I look around for the gardens are to be landscaped.

A northern family passing through. They have ordered a mixture of coffee and tea, and Big Fat Boy breakfasts. The two women are large. One of them, Mum, bears a passing similarity to William Perry, who was nicknamed 'the refrigerator' and played American football some years ago. Whatever happened to him? Mum's son, sunglasses resting on his forehead, surveys his breakfast with the glee of a young child. Mum puts sauce on his plate. He is in his late twenties. Complaints about the price of the food are audible as Dad rolls along to purchase yet more food.

An Eastern European man moves cautiously around the food hall, skilfully avoiding eye contact as he negotiates the cups and saucers counter. He is joined by a compatriot. Both have olive-colour skin and short dark hair, probably cut at home. They wear blue bomber jackets and jeans you get from the ever increasing numbers of high street shops where every item is on sale for £1. Conversation is muted and controlled. Glances come over in my direction. They catch me looking at them.

Can't keep my eyes off a woman in a peach-coloured skirt with matching white blouse. Her dark stockings are baggy at the ankles. She has a 'business suit' with her with immaculate creases in the trousers but no tie. They sit and laugh together. Both lean forward on the table to catch each other's every word. Smiles are exchanged. She sits back in her chair viewing 'her' man. She's blind! Her guide dog rests on the floor beneath the table licking its bum.

29 September

It's Saturday afternoon. Sit with a good friend of mine, the pre-octogenarian, John – ex-Spitfire pilot in the last war. It's his first time here.

'It's bright in here. It's the sun, not like when I was a boy. It's the hole in the ozone layer.'

I nod.

'Where do we get the coffee and tea?'

'What, no cups and saucers?'

'Over there, John.'

He squints at two women loading cups, saucers, and cartons of half cream, spoons and napkins on to trays.

'Bet there's two inches of dust up there.'

He's glancing up at the roof. Talk is about how the steel butts and roof are cleaned.

'Did you see the price of the coffee and tea?'

'Staff are all bloody foreigners?'

'Why's that?'

Two elderly women in twin set and pearls join us. Their pleated past-the-knee skirts, black shoes with gold buckles, and white blouses match their seriously plummy voices. John nudges me. They are joined by a Church of England vicar carrying a tray of teas. His white hair is parted on top by a bald patch; he has Harry Enfield's teeth. His Evensong voice reaches over to us clearly. I am dazzled by his purple-coloured top and white dog collar. The three of them speak in unison as tea is served impeccably. The women's legs gracefully hide beneath the table, whilst the vicar smiles at all and sundry in between sips of tea.

John is mesmerised. We slip out quietly as the choir begins tuning up and unsuspecting customers are guided to their seats for the service.

'Is it always like this?' he asks me as we hunt for his old blue Saab in the car park.

1 October

It's Monday morning. There's a shortage of staff this morning. Chris is busy hoovering the floor behind the counter. A pleasant African woman serves the coffee and teas with no hot water. The Red Hen table service is closed.

A group of 'grown-ups' sit alongside me. Much talk of travelling. I pick up a mixture of Jamaican and English accents. Five men and two women; one of them is very sun-tanned. Coffee, tea, water and orange juice flow. A white guy appears and the group moves off with another four white women. They all climb into an unmarked grubby white minibus with a missing back registration plate.

A man in a blue Barbour Gamefair jacket with John Lennon glasses and a blue prep school satchel, asks one of the table clearers if there is a menu available. He receives a bemused look and some garbled English.

Some Portuguese is hurled across the tables. A colleague quickly appears with a blue J-cloth. One of the house rules has been broken! Speaking to the punters is not allowed. A suited Welcome Break manager is summoned and intervenes politely informing the customer that he has to order from one of the counters. Order is restored, the 'offending' table clearer is asked to display his ID card more clearly. Smiles all round.

I connect with a 'dodgy' geezer in a faded blue denim jacket, with matching dark blue shirt, and topped with a yellow and black polka dot tie. He walks in with his hands in his pockets, and a Filofax tucked under one arm. His black creased trousers hang limply over scuffed shoes. He watches a blond woman leave, his glasses glued to her wriggly arse. He sits with an open-necked ex-boxer wearing a middle-aged man's black leather jacket picked up from the Hackney Wick Sunday market. They sit between a suited businessman. There is much shuffling of feet and they are joined by another man suited and booted. The four leave together knocking incoming customers out of the way.

*

The location of toilets affects spending. Consideration is given to placing these facilities at the back of the station to draw customers through the mall.

*

2 October

I am caught up in a scrum of little old ladies, all four feet nothing, with small twisted white curly hair and misshapen toes poking through their M and S sandals. Pearl earrings hang over sunken cheeks.

'Where are you going, Ethel?'

Her friend guides her back in the direction of the chairs they have captured. Danish pastries are eagerly consumed. Teacups are held permanently three inches away from mouths. Tea is their lifeline. With no men around to hinder their talk, dentures are regularly cleaned by a sweeping motion of the tongue. An overfill of tea results in Gladys rushing off to the ladies'. Her friends watch her disappear. Black handbags under the tables are waiting patiently for their owners. Ethel is dozing at her table.

'All right?'

A vacant nod. Her friend in a blue trouser suit washes the remains of a scone down with a drop of tea in between nodding and smiling.

'You're all right then?'

Barbara with the collapsible walking stick is called back into the fold.

'It's okay, I know where I was going.'

Freshly watered and fed, the scrum slowly disperses to explore the shop. A pungent smell of rotting cabbage is left behind as Terry the coach driver stands waving at them from the exit to hurry up.

*

It was clear that the majority of travellers use motorway service areas infrequently. Three quarters (75%) of respondents used motorway service stations less often than once a month. The majority (83%) had used the toilets, over half (57%) had bought food and drink in a restaurant or café. (Highways Agency Road User Satisfaction Survey)

*

3 October

It's early evening and the car park is full of white transit vans. Two middle-aged white 'spam heads' stand peering through the back window of one of them.

I am sure some people become like the 'magic Christmas trees' hanging in their vehicles. You know, like people over time begin to look like their pet dogs.

It's been a long day and I grab a bottle of water and *The Indie* from the shop. The *Guardian*s are all sold out. I am the only customer apart from two lads perusing the mobile phone stands. The two African women on the cash tills look bored out of their heads.

I manage to avoid a giggly group of sixty-year-old somethings with delicately manicured fingernails firmly gripping polystyrene cups. They sit six deep to a table. A mixture of M and S and British Home Stores clothes. Their talk is of wayward husbands, grandchildren and holidays.

A 'midnight cowboy' fellow walks across my path, carrying a newspaper and a bottle of champagne. He ties up his 'horse' at the Burger King corral, his denim suit and brown cattle – punching hat shine as he orders his meal. He saddles up and heads off towards the Days Inn accommodation with the stride of expectation.

The smell of steak and kidney pie fills my nostrils. Being a signed-up vegetarian of some years standing, I grit my teeth and imagine I am somewhere else, to no avail. The carrots and boiled potatoes with gravy enhance the aroma. Feeling a bit queasy I gulp at my bottle of water, managing to tip it down my shirt. Two women walking past, the remnants of the 'manicured mob', look at me with pity. I smile and bury my head in *The Indie*.

4 October

'He's got the fucking key. Why can't he open up?'
 'It's what I've been telling you.'
 'He can't do it.'
 'His company, his money.'

I arrive at the end of a mobile phone conversation.

The man in an olive-green shirt with a matching dark tie, and carefully groomed jet black hair, speaks directly at the man sitting across the table to him and down his mobile simultaneously. The second man in a dark off-the-rail Burton's discount suit is quiet.

The 'mouthy' one reminds me of a younger Dave Bassett when he was manager at Wimbledon. You know, always up for it, talks bollocks, and is oblivious to those around him.

Lots of arm waving and 'fuckings' as they leave.

Sometimes I feel as if I am on the set of the *The Truman Show* with Jim Carey. Is this an artificial world I am in?

*

Motorway service areas get such a bad press that many motorists do not stop at them preferring instead to drive non-stop to their destination. Guidebooks such as Off the Motorway *by Christopher Pick have been written to provide attractive alternatives for a 'tea and a pee' off the motorway.*

*

7 October

It's a black, windswept early evening. Rain-sodden clouds are racing across the sky, dumping their contents on to the busy car park. The place is alive with loads of kids in David Beckham shirts – well, he did score a third minute injury time goal yesterday in the World Cup qualifier against Greece. When I look around, seems every male over the age of five is wearing a Manchester United football shirt.

Barry, complete with black baseball cap and nervous hands, finally serves me at the twenty-one person length queue. I fight off an elderly woman wearing ankle stockings and a grubby blue raincoat to grab the last few mini milk cartons. Her dark long grey hair covers her eyes. A retired professor and his wife, wearing a headband, bump into Mr and Mrs Country Life at Burger King.

'I've waited fifteen minutes for these chips.'

A large man wearing a Wales rugby shirt stuffs chips into his mouth oblivious to the fact that a harassed mum with two crying kids is still waiting for her chips. With Burger King several people deep waiting, tempers are becoming strained.

A suspicious man in sandals with socks – he must be English – walks off in disgust towards the direction of the gents'. I wonder what he is carrying in that green, over-the-shoulder camera bag?

Three Africans are busily purchasing clothes at the 'Official Grand Prix' merchandise outlet. One of the African security guard stands next to them at the cashier's till watching passively and muttering in French to them.

I become conscious of a man staring at me. He looks a bit like Chevy Chase, in his role in the film *Spies Like Us*. He has that cheeky 'I know what you are up to' smirk about him.

Never seen so many young children here – all shapes and sizes with some competing as to who can make the most noise and be really, really irritating. Others sit mesmerised by the food before them; others stare at what is happening around them. A little girl picks her nose with her 'chips' fingers.

With the hanging televisions not working the place sounds like a cross between a busy pub on a Friday night and a large open-plan office which has turned itself into the local church playgroup.

A baby is being fed by his very serious-looking mum. In between spoonfuls of a brown custard substance he watches his three-year-old brother play with his chips which are lined up in a row on the table, with his burger split in many parts alongside. Dad, in a Levi's jacket, holds an older brother who has kindly thrown up on his denim shoulder. It smells of cheesy milk!

A group of young Malaysian women turns up. All are dressed in black leather jackets and jeans, carry bulging white plastic bags and bottles of water. The oldest looks nineteen. They sit alone and draw a few stares from the Muslim couple, heavy with Leicester accents, who sit alongside them. I am reading Kanan Makiya's article in the *Sunday Independent* about the challenge for Muslims the world over since September 11 and the destruction of the New York World Trade Towers. Words are being said out of reach of the couple, but they respond by getting up and leaving. I feel embarrassed at the ignorance of some people.

A mega-rich woman with short blond hair appears with her tray. She sits alone quietly talking into her mobile phone. She has more 'serious' designer labels on her than you see in the private boxes at a Premiership football league ground.

Three young children run excitedly past my table. One gets a clip around the back of the head by his mother. He immediately sits down and mauls a cold burger. His two 'new' friends roar off in the direction of the empty smoking section.

A family with three teenage girls, all with some degree of learning disability, troops out. The younger children of the family skip merrily ahead. Mrs 'Mega-Rich' follows them out.

A father, middle class but with a Thames Estuary accent (or is it Mockney?), grapples with his daughter who is refusing to get up off the floor. The tomato sauce down her gingham dress gives the impression she has been stabbed. Mother gives approving nods as he lifts her up to much screaming and kicking. The family, complete with elderly embarrassed aunt, leaves to smiles and glances from other parents.

I've lost it.

A woman with weightlifter's arms, and short ginger hair with a purple headband and wearing a strapless dress, sits with five other women. Her purple and black coat rests over the chair behind her. A bag, which is black with psychedelic orange splashes on it, sits alongside her. She has a snake tattoo on her left shoulder. Her silver shoes hang limply from her feet. A friend opposite with bright blond hair carefully puts lipstick on in between taking sips from a can of Coke.

It's raining now outside but people continue to emerge from the car park in various states of being under-dressed in shorts and tee shirts. Others enter wearing an assortment of mountain and high altitude fleeces, long Siberian coats complete with hats bearing ear-flaps, scarves, and waterproof jackets. I think I spotted one woman wearing an all-in-one wet suit?

By eight o'clock there are less people around. Families with children are departing. Grown-ups are coming in along with pairs of teenagers with 'love' in their eyes, holding hands. Three elderly grey-haired couples walk in blinking as the confusion of colour and light hits their eyes. The Malaysian group exits in single file like men from the small islands of Western Ireland.

A small baby dressed in a blue Mothercare babygrow appears on the floor beside me. There is no parent in sight. Not sure if it's a boy or girl. A milky fluid pops out of its mouth on to the carpeted floor, which smells like an outside men's toilet of an East End pub. I vacate my table.

Bump into Alan Bennett (or is it?) as he crosses my path en route in the car park to my car, his hands firmly clasped behind his back, wearing a zip-fronted jacket and a pair of Ted Heath yachting pumps. His black glasses bob up and down as if he's searching for something.

8 October

Two very sun-tanned people tuck into a Fat Boy breakfasts with extra portions. He reads the newspapers whilst she concentrates on eating as much richly buttered toast as possible. Conversation between the two is muted. They look as if they have recently arrived back in the UK from the Costa del Sun and driven straight from Heathrow Airport around the M25 for an English breakfast. Fifteen minutes later still no conversation between the two. He eats and reads; she is keeping her head down. The man dressed in a card player's black shirt, with a couple of silvery pens in the top pocket, retrieves some more cartons of milk from the counter. His white jeans fit snugly into expensive brown casual slip-on shoes with no socks. They settle down again and return to their newspapers.

Dave takes a phone message on his mobile.

'Dave speaking.'

'Queuing up for my breakfast.'

'About fifteen minutes.'

'Could be worse.'

His other hand is putting his signature to a credit card receipt. The rest of us in the queue behind shuffle our feet. With bacon and egg roll in one hand and his mobile in the other, he jogs out.

Welcome Break TV starts up in my left ear to the sound of grandma in her wheelchair chewing her toast slowly and deliberately alongside me. Her granddaughter, in wet-look crinkly hair, hovers around her. Her husband dressed in 'going away' clothes, watches in between drinking his coffee. Grandma's cheeks puff out like a hamster. Her granddaughter passes over a napkin. Some food is returned and the napkin is placed on the tray discreetly. Grandma is wheeled to the toilet. Thirty-five minutes later they return. Her ulcerated legs are a purple freckled colour. She clasps two bunches of flowers fresh from the shop. The husband hobbles after them, clutching his NHS collapsible walking stick. I notice the granddaughter waddles on worn out hips. They all leave together, laughing as they exit.

*

Please excuse our appearance during this refurbishment to provide you with a bigger, better, bright Break.

*

9 October

Text message from Jim.
 <Breakfast at South Mimms... mistake... Bombay wedding meets Baden Powell jamboree with ongoing building works... what a racket!>

10 October

It's just after 8 a.m. and two dazed-looking people are gazing longingly at the closed Burger King outlet. The man, with dyed blond hair in a ponytail, and wearing a Kangol jacket which is tight around his neck, stands staring disbelievingly. His hands fidget in his pockets. His woman friend, blond hair in dreadlocks watches with arms folded.

'Yes, it is closed,' I hear myself saying in my head. They still don't quite get the message. The lights are off, one member of staff is busying himself cleaning up in the back.

8.08 a.m. The couple retreat to The Granary with hopeful heads occasionally turning back in the direction of Burger King.

A young workman in blue overalls slinks past carrying a litre bottle of water. His skinhead haircut reflects the sunlight from the glass roof. This 'halo' effect lasts momentarily as he asks his mate in a South London drawl, 'What's happening now then?'

'Fuck knows!' comes the reply.

'When's he coming?'

They disappear behind the newly erected screen which hides

what is going on inside the closed Game Zone. Another sign warns that it is a hard head (or is it hard *hat*?) area.

Three men cross my path. One is in a dark pinstriped suit which covers a creased blue shirt. A younger man in an open-necked shirt has gelled spiky middle-aged hair. The third man sports a pink striped shirt, with an off-the-rail Next tie. He speaks with a heavy Welsh valley accent. Papers are spread over the table. The air is punctuated with insider business speak: 'short on targets', 'manoeuvres', 'images', 'meeting budgetary constraints', 'bring them back to budget at divisional level'. The presenter, Mr Spiky Haircut, fires off words in rapid succession. 'Converting pretty strongly at the moment', 'no thrills', 'back to under 4%'. It all feels a bit like a Barnum and Bailey circus performance.

*

The introduction of motorways in the 1960s came with waitress service and silver cutlery.

*

11 October

Go for my first breakfast at the Red Hen.

'Smoking or non-smoking, sir?'

Isn't it nice to get called 'sir' occasionally?

'Non-smoking, please.'

'Certainly, sir.'

I am shown to a table and the menu placed under my nose.

I sit next to two businessmen who work for a chain of commercial pubs. Both are tucking into the full £7.99 breakfast complete with a liberal helping of flatulence-inducing baked beans which, looking around, a number of people seemed to have left on their plates.

A mixture of other men sit around. All read newspapers. It has that 'gentleman's club' feel to it as Natalie brings me my coffee. People walking past the Red Hen peer over the barrier separating us from them. Do I stare at the eaters as well when I am passing?

'Could I have the traditional breakfast without the meat, please.'

'Certainly, it's quite a popular request.'

Not this morning, evidently, as half-eaten sausages sit on plates.

Natalie beams back at me as she takes my order. It feels a bit like home as I eat. I read my newspaper and have that warm glow inside which always comes when I am feeling good and the HP brown sauce is flowing. Natalie periodically pops back to check if I am eating it all up like a good boy. I eat my beans with some relish; they're good.

Two lads from Lancashire sit across from me mulling over Manchester United's easy win last night in the Champion's League match against Olympiakos at the Olympic Stadium in Athens. Being a Tottenham supporter, my memories of away victories in a European Cup match have somewhat faded.

Leaving, I fart loudly as Natalie wishes me a good day.

7: 'Hello Tel'

Unless you actually know the landscape my diary photographed, you've no option but to accept my version.

Ernesto Che Guevara

Gothics – Hi Man – Timothy's little holiday – Hilda and her three friends – on the gaming machines – public school chaps – Mrs Neville's cycling days – Jim's lost mobile – Russian money – Mr Taylor's nursery days – two country gents – mobiles – Mr Big – a confession – opera in the gents' – my dad's greenhouse.

13 October

I seriously fancy a Saturday afternoon Danish pastry and a filter coffee.

'Sorry, sir, we have no plates – they are being washed up'

'Yes, but when will they be dry?'

A quizzical look.

'I am afraid we have no filter coffee, sir; will ordinary coffee do?'

For £3.35 I get an ordinary coffee in a tarnished silver pot complete with a lid that doesn't close properly, and a Danish pastry on a napkin. I retreat to my seat.

An elderly woman, with a hook nose, slurps noisily on a straw from a Coca-Cola tub. Her grey hair is combed forward and reaches the shoulders of her Hawaiian style shirt with its multiple colours and no design. Her earrings dangle down amongst the folds of aged fat. The two people she sits with are bearded and bored. Both stare aimlessly into space. I am intrigued by one of companions: a man with a dark blue blazer, with a rolled neck top, white clean baseball boots, and baggy white trousers, all of which match his white beard. He looks like a Cowes Regatta yachtsman without his boat.

Catch sight of an average-sized young white guy who is walking around the place in Levi's jeans and trainers and little else, bar a gold chain around his neck. Two women stare and giggle. He is shouting and waving but no one pays any attention to him.

A group of four 'Gothics', all with long, dank black hair, regulation black clothes and black boots, look menacing – apart from one, probably a male, with his metal crutches lying under the table. Their tight jeans look welded to their seats. A Mr and Mrs 'Weekend Large Newspaper' readers at the next table pretend not to notice them and hope they are ignored in turn.

Mr Bare Chest is hovering around the 'Van Gogh's Work Room' and is attracting the attention of a security guard. He watches from a discreet distance, one hand in his pocket, the other hanging limply by his side. He has an unruly young boy with him. He is coming my way to the table next to me, where a bald, goatee-beard fast eater is busily tucking into pizza and chips. They have travellers' accents. The little boy is shouting 'Daddy' at the top of his voice.

'Daddy, I want to talk to you!'

'Shut up, I said.'

The boy appears not to notice the swipe around his head.

The 'Gothics' are clustered around the mega grab machine excitedly shouting encouragement to the woman who has won a lime green fluffy animal. A man, with a face like a barking dog and wearing truck driver's boots with matching bright yellow trousers aggressively bumps into the face of a Muslim woman pushing her elderly mother in a wheelchair. Both hide their faces under white headscarves and make their way to the toilets. No apologies are offered.

The travellers' insistence on returning to the Van Gogh Work Room has forced the security guard to return along with one of the 'day' managers, the Keith Vaz MP lookalike.

In the car park an elderly Sikh, with a black turban and a long white beard, is sitting in a new Ford car with a younger woman. He leans out and spits on the floor, narrowly missing my right shoe.

Mr Bare Chest Man races past me as I walk to my car. He's

driving a beat-up old black Volvo, one of the 3 series, similar to the one I sold to my mate Jim a few years ago. No seat belts. I watch him and the boy screech to a halt near the exit. They disappear into the early Saturday evening motorway traffic.

15 October

It's early. The place is empty. I kick a tumbleweed out the way as I walk across the deserted car park.

Inside, a security guard stands looking at the BT Multipoint Email Link. A man with a mobile phone walks up and down talking to a business colleague, his West Country accent is loud and harsh this time of morning. Three open-necked young computer businessmen sit relaxed sipping their first lattés of the day and talking.

A fat man in a black shirt, rumpled trousers and spiky beard, stands far off staring into space. He has a David Blunkett smile and a waistband the size of an old red routemaster London bus. He waddles off to the gents' causing one of the small Welcome Break staff to jump nervously out of his path. He reappears later holding two takeaway bacon paninis and a bucket of tea. His white plastic carrier bag bulges with food provisions for the next leg of his journey.

Before leaving he plays the machines in the Game Zone. His mobile phone is strapped to his giant waistband. He catches me looking at him and gives me a half smile through dirty teeth.

A young woman in her early twenties, but dressed older, is sitting reading Peter Abrahams' book *Perfect Crime*. She picks bacon roll bits from her teeth as she reads. Dressed in a black trouser suit, she seems to be waiting for someone. Carefully she avoids eye contact with people around her. Her mousy-coloured hair tumbles over her reading glasses. An expensive diamond ring sparkles on the middle finger of her left hand. He appears. She disappears into his embrace – the arms of an older man, suited and suave, greying hair, with a nice red tie which matches his socks. They leave separately in new leased cars.

The 'big cheese' manager Stuart is out and about on the shop floor. Staff busily polish counters. One of the managers is re-arranging the fruit on The Granary counter before moving on to the HP brown sauce sachets. Stuart walks around inspecting his domain, mobile phone constantly in use. An underling, an older man, hovers around his boss.

Later...
'I can save your company £170,000 per annum.'
Three people cluster around looking at him in bewilderment.
'But how about increasing staff costs?'
Casually dressed, in designer label sweatshirt, hair like an ageing Seventies rock band singer he smiles and says simply, 'Reduce overheads.'
'Could I finish my tea?' asks the only woman present, who sports an impressive company name badge on her jacket lapel. Voices are lowered, heads come together over the £6.99 Ultimate Breakfast leftovers.

Is there a time of day when men are more sexually aroused than other times of the day? I only ask because the sight of a thirty-something woman walking through with her coffee is enough to turn a few men's heads. Okay, if I am honest, mine as well. Is it her shoulder-length blond hair? Her tight-fitting black trouser suit? The leopard skin type blouse? What? Eyes sneak a view over a laptop computer. Fingers twitch in holding the business pages of newspapers. Others just stare honestly. Once stirred, the surrounding men take time to settle down back to their other 'duties'. I detect an audible rise in the conversation, newspaper pages rustling and fingers again tapping computer keyboards.

18 October

Bump into a 'hippy' – a 'real' traveller. Got to be aged fifty-plus in his rimless glasses. His wiry frame is covered by a light-orange mountain Afghan type coat over a multi-coloured lumberjack jacket. His trouser pockets bulge with hidden mysteries. He sits with his coffee and cake surveying the other 'punters'. Feet rest on a chair, brown boots which have seen 'action' up an Indian magic mountain trail. He has that faraway 1000-metre stare which I recognise in myself at times, particularly when surrounded by the meaningless chatter and 'in your face' banality of western consumerism. He has had enough and departs with a sage-like expression. I catch sight of his old battered VW camper van pulling noisily out of the car park, which is now stuffed full of shiny new cars.

20 October

Timothy's 'little holiday'

'It must have been like early summer because by eleven it was dark. I spent a Friday evening there. I got there because I put diesel in a petrol car by mistake. So I rang up the RAC and after a long rigmarole they finally found me on the motorway and said it had to be changed. They could take me to South Mimms. So that was good. Told me it would probably take hours. So I got there about seven.

'So I went to the public world of the service station. What was odd about it was how quickly, within half an hour, you felt you shouldn't be there. Considering it's a public space and there is so much going on, it's actually designed for eating and going. To be there longer than that – I think that is how I coped with my making eye contact with somebody; it just actually established the fact that I was a legitimate person, not a loiterer and the place is designed for loitering, but somehow you can't be a loiterer there, you have to be doing, you have to have a reason for being there, you have to have a purpose.

'I actually got up a relationship with the people behind the counter. It's really strange because I got a couple of coffees and what I was really impressed with was the people behind the counter were such a mixed bunch; most of them didn't seem to be English. The English person who was there wasn't the brightest but she was very efficient and very nice, and very helpful. Helpful in the way of actually seeing people who say they are having difficulties. Helpful, like I've been trained and this is what you do. She was really nice. The care and consideration they took and the horrible job, you wouldn't bother.

'Before me there was a French family come in and they wanted the breakfast and she explained in precise detail why it was actually cheaper for them to have the all-day five item breakfast rather than buy three items. They just could not understand, but she explained it in this very meticulous way and kept going over and in the end they, I think they were more frightened than anything else, just agreed and paid. You could see them sitting a bit further down trying to work it all out.

'The people who come round and clear up – they were very polite. You somehow become part of the furniture after a while, part of their scene; you have entered their little world for that hour and a half. From there on I bought some sweets and magazines and felt at home. I also bought a jacket from the shop, so it was almost like having a little holiday for the night in South Mimms, you know!'

21 October

It's a wet Sunday morning. Churches are empty. Coaches are lined up in the coach park like beached whales. Children dash from the toilets to cars through the pouring rain.

Inside I am greeted by two Hindu women venturing out into the monsoon rain wearing sandals with their toenails painted red. Two policemen eye me suspiciously as I look over my cup of coffee to what is happening around me. Two men with short hair, one without shoulders, lost under his Hugo Boss jacket's collar, avoid eye contact with the two policemen. They both smile as the two police officers don their yellow fluorescent jackets and move off.

A woman in her sixties feeds her husband sitting alongside her. He has a death mask on – white face, sunken cheeks – and with wobbly hands he smiles lovingly at his wife as she sorts out his teacup and pours his tea. She is talking into his left ear as he munches a cake. Occasionally nodding, his open mouth is like a deep black mine shaft. As bits of cake gurgle out from his mouth, his wife lovingly mops up the spillage. They hold hands under the table. I am caught looking at them; they both smile at me. I smile back and want to go over and hug them both.

I am overcome by Hilda and her three friends' perfume; it has swept over me. My glasses become steamy.

'I got up at half past five for a wee and stayed up.'

'I done that the other day when I got up for the hairdresser's, had a cup of coffee and stayed up.'

'How much was that?'

The tea has arrived. Curly grey heads bob in a rhythmic chorus as sips of tea are taken.

'I just want to pour you a cup out, that's all, love.'

'Nice cake, what is it?'

More cartons of milk arrive, courtesy of Peggy.

'Can you taste it?'

'That's half cream.'

'Is it semi-skimmed?'

'Marvellous what they can do with milk now.'

'Have you seen that advert on the telly?'

'Have one, go on, go and get one.'

'What time does the coach leave?'

22 October

Group of young men, a cross between young footballers with a professional football club, trainee police cadets, and young lads passing through en route to a party confront me. Most wear designer sweatshirts. They are re-fuelling on Fat Boy breakfasts, Cokes, and bacon and egg paninis. Newspapers start appearing, not the *Sun* but *The Times*!

One of the group asks to borrow a spoon off another table. Another in a hooded sweatshirt, who would be mistaken for a 'druggie' in Hackney and given a wide berth, drinks his strawberry Fanta with public school breeding. Wearing an All Blacks rugby shirt, the short ginger-haired bespectacled member of the group enquires as to the 'state of play'. A couple have black eyes and puffy faces. The only black member of the group is looking through the *Guardian* at Saturday's football results. They are impeccably behaved, no sign of a teacher with the group apart from a track-suited Will Self lookalike who only appears marginally older than his charges. A large bottle of Barr sits quietly on their table.

23 October

'No way, man!'

I look up to face a small woman wearing a green Nike baseball cap playing the 'Cops 'n' Robbers' machine next to me. She screams and yelps. At 30p a go to get the 'swag', she could be here some time.

The flashing coloured lights from the 'Crazy Fruits' machine I am playing are making me dizzy. Can't seem to get the fucking three smiling bells to win. I've put in £7.50 so far with no result.

Move over to the 'Hot Stuff' machine where its scantily dressed woman successfully relieves me of £4 in less than four minutes. I am having problems with the instructions. 'Hold', 'Nudge', 'Control', 'Active', 'Start'. Where's 'Win'?

My two co-players both look under eighteen years of age. One of them, a spotty lad, seems to be losing heavily on the 'Royal Roulette' with its tempting £25 first prize.

As we play, people periodically look in as they pass by but do not enter.

£11.50 down I give up and to loud constant laughter from one of the machines, head off for a strong coffee.

24 October

Walk into the toilet to find two men standing around. A nearby fire alarm is ringing incessantly.

'What's up?' I ask.

A trucker in a red check over-jacket, brown boots and dirty jeans, looks me up and down.

'They're all full.'

His hair, blond going on grey, covers his face, and a beard is growing nicely through the lack of direct sunlight. The other man, dressed in a regulation company blue driver's jacket and trousers, dashes off to the far end to a vacated cubicle.

I stand and stare at the wall counting the white wall tiles as I await my turn.

Later...

Decide on two eggs on toast this morning. You know it's one of those days when the smell of HP brown sauce is overwhelming. Unfortunately, there's a shortage of seats. I end up sitting under the 'This is live satellite television from Welcome Break' television screen. My early morning brain is being fried.

Between mouthfuls of grub I watch a clutch of beautiful business-suited women all carrying off trays loaded with coffee and fruit juice towards the direction of the smoking area. Too late, my eyes fall upon a woman in black stockings, short tight skirt, and small black bootees. Her long blond hair is falling out of a bandana. I quickly take my early morning PC pill and retreat to my newspaper.

A group of soldiers enter and immediately search out an early morning Burger King. They are short on local knowledge – it's in darkness. Seeing soldiers in battle fatigues, carrying their light-blue berets and well-worn polished black boots, always makes me nervous. Like, where do they keep their guns? Are they safely stashed away on their vehicle or discreetly hidden beneath their baggy trousers or jackets? One of the group, a short blond woman, hair up and tied behind, looks hard. She has an unemotional stare in her eyes.

The Sikh businessman in a light fawn suit and black turban ignores them, reads his newspaper and slowly sips his tea. Facing him is a middle-aged white man, with greying hair, early fifties, and dressed casually in blues and blacks. He sits hand in hand with a small Filipino woman. She is half his size and wears new Adidas trainers and baggy training bottoms.

25 October

Two businessmen with takeaway coffees.

'I do love those black blank screens.'

'Yeah, £5,000 a piece.'

Welcome Break TV has not woken up yet!

The place is quiet, bit like a dentist's waiting room; we all seem to be facing in towards the centre. People stare at each other. A large family to my right provides the only entertainment with their talking and laughter.

'Good morning, Mrs Ryan. How are you?'

'Slightly bleary-eyed.'

A middle-aged white *Daily Telegraph* reader exits with a large posse into the early morning darkness.

Can't help noticing a family of four. Man, woman and two pre-pubescent boys. Mum is dressed in all denim, whilst the man is more upmarket in casual wear in a roll-neck sweater. The adults kiss and make 'cooing' noises when the boys go off to the shop to buy handheld computer games. Mum and 'Dad' exchange passports and laugh at each other's photographs. Mum disappears to the toilet leaving 'Dad' to make small talk to the boys. It's painful to watch and listen. He so wants to impress the brothers with explaining how the game functions, but they refuse him a go. The three sit in a line. Dad gives up and sits in silence staring out into space. The boys continue punching buttons on the toys, while Dad fingers his mobile phone.

Mr Biggest Belly in the World has appeared. He sits two seats away from the table he shares with two companions to accommodate his beer-barrel sized belly. It's not pleasant watching fat people eat a double beef burger, with trimmings, first thing in the morning.

Mum has returned and takes over the mobile phone from Dad. She reaches across the table to hold Dad's hands and leans back in her seat; alternatively the boys watch bemused at their Mum's antics. Mum and Dad leave the boys and walk off hand in hand. Dad caresses her left buttock as they look at the flower stand. The brothers' stares follow them.

28 October

Mrs Neville is 71. She used to live in Potters Bar and now lives in Spaulding in Lincolnshire.

'I last went to the service station two or three years ago when I had to go to London. I travelled down in a snowstorm and stopped there. Appalling. Awful place. I am not gone on motorway stations anyway. I had a coffee there. I thought the place was garish, with its concessions. That was the impression I got. I didn't think much of it.

'Back in 1949, early 1950s, we used to meet at the road transport café, The Beacon, which was there before it all changed and they built the motorway and the services. It was a lorry drivers' café. We drank tea there and ate their revolting breakfasts. It was grot personified. My husband stayed there once, not sure why. I know a few ladies of the night congregated there although I didn't know what that meant at the time.

'We used to get there before six o'clock in the morning, often at 5.30 because the police wanted the whole thing over by nine o'clock. There would be over a 1000 cyclists from all different clubs, including marshals, spectators and hangers on; it was a good crowd. The whole yard would be full of cyclists.

'The cyclists went off in one-minute intervals for the 25-mile time trials. The course was up the Barnet Bypass although we called it the Arterial Road. It was a two-way road then.

'We had pukka racing bikes – none of your mountain bikes. Lightweight bikes with 27" wheels. Fixed wheel with a 72" gear. Some people had speed gears. There were Claude Butlers, Bertrams, and Holdworths, all handmade to your size. You had to go to a fitting to make sure you got the right size, just like for a suit in the old days at Burton's.

'Our club was the North London Branch of the National Cyclists Union and NORION was the affiliated cycling section. There were other clubs, like the Barnet Road Club and Tottenham Phoenix.

'After races we sometimes went in the Middlesex Arms pub. Awful place. Huge cold empty place with long trestle tables and benches. Not very welcoming. They wouldn't let you stand at the bar. You had to sit down with your drink.

'This was in the late 1940s, early 1950s. There weren't many cars around nor juggernauts – not like now.'

*

KFC: A fast food outlet specialising in fried chicken, renowned to be 'finger lickin' good'. Exclusive to Welcome Break on the Motorway.

*

29 October

The clocks have gone back, summer has ended.
 'I love you, too.'
 'No, I am sure I left it by the cooker.'
 'Not too bad.'
 'Probably about seven.'
 'Okay.'
 'Bye.'
I am sitting in cubicle 5 in the gents' reading my *Guardian*.

Later...

A bloke reading his *Sun* newspaper sits across from his spindly wife. Tits and bums face her as they hold a conversation. She has two gold earrings which almost stretch down the length of her neck. He looks the type of man who is on first name terms with the local garden centre staff.

Aston Villa are top of the Premiership, and British Muslims are being killed in Afghanistan, whilst women and children have been gunned down by Islamic gunmen whilst praying at their usual Sunday morning service in Bahawalpur, Pakistan.

A couple of Liverpudlian builders passing through are joking about the coffee they are drinking whilst a coachload of over-sixties 'grey heads' have arrived and are causing confusion by joining the queue for tea and coffee at the wrong end.

My mobile phone rings as I am caught up in a conversation with two businessmen who have pulled me into a discussion they are having about motorways.

'Hello?'

'Yes?'

'This is St Albans police station here, nothing to worry about, sir. Someone has lost a mobile phone and your name is on it.'

It's my mate Jim's phone. It seems he has dropped it somewhere in the Hertfordshire countryside and it's been handed in to the police.

'Give me your number and I will get him to ring you.'

'It's Dawn, I am on the front desk at St Albans Police Station.'

A few smiles around me. People have been listening to my conversation with Dawn, despite the early morning 'noise' coming out of the Welcome Break TV screens and a crying baby.

*

'Junction 23 to 25 South Mimms westbound traffic is slow and building. Thank you, Keith, for that phone call.' (Radio 5 Live Traffic News)

*

30 October

The man with Tourette's syndrome is back. Sitting alone he stares into space, hands clasped together, occasionally glancing up at the TV screen. Coffee is sipped delicately from his cup. It's as if he is waiting for something, or someone. He appears restless. People sitting nearby are not in tune with him. He checks his watch, carefully finishes his coffee and leaves. I watch him depart – a slow casual walk to the exit punctuated by the occasional bark.

31 October

A male twosome for lunch. One keen on cleaning his ears with a pencil wears second-hand army gear with black dirty army boots. His hair is cut short with a small braided ponytail and assorted ear studs. His companion has on his head a mousy coloured wig, which hangs over black glasses, and a moustache which has an unnatural tint to it. He is dressed in different shades of green (jacket, shirt, tie and trousers), a bit like an overgrown pixie. They both look and feel vaguely Bavarian but speak in fluent Black Country English.

2 November

The 'boss' is showing other suited businessmen around the site. I keep my head down over a plate of eggs and toast, washed down with a pot of vintage Welcome Break coffee. Staff stand briskly to attention as he moves imperiously past them.

The 'funeral director' sitting opposite me sees this and then looks me up and down as if measuring me up for future reference.

I settle back down to reading my paper and try to avoid squeezing the chocolate out of my freshly baked croissant down the front of my tie. It's a serious tie today because I have one of those meetings later on where those present seem to be more concerned about your dress than what you are saying.

My attention is drawn to two bare-chested lads with slightly tanned backs and striped sports trousers sitting across from an older man. I watch him count out three neat piles of used £10 and £20 bank notes. The accents seem to be Russian, the style, dodgy East London. It's ten past eight in the morning.

5 November

Mr Taylor is in his late seventies. We talk in the front room of his house in Borehamwood. Mrs Taylor serves us tea and biscuits. He used to work at the nursery at Bignalls Corner before the motorway and the service station came.

'The nursery was on the crossroads that existed then before they built all that enormous great motorway system. It was on the south side of the crossroads, the Six Bells was on the north side, Bignalls Garage – they had two or three garages as well as the nursery – was on the west side, and the Middlesex Arms Pub was on the north side. That was the crossing. It was quite busy. There was a big café which I think is still there; has it gone? There was the Beacon Café that was there – that was a pull-up for lorry drivers; they used to come into London and stay there overnight, then travel on into London the next day. They came from all over the north, Scotland, the North-west, Manchester, Liverpool, Sheffield, Nottingham, everywhere.

'The lorry drivers would come along and they would be dropping girls off and would be picking others up. "Old toms" as we used to call them. I was a young lad and didn't know much about this sort of thing but these women were working the road. They would be standing on the roadside waiting for a lorry to come along and pick them up and off they would go and, of course, they used to use the Beacon Café; they used to pull up there overnight sometimes. It was like a brothel there. As a young lad I didn't realise it because I was an innocent sixteen- or seventeen-year-old and in those days you didn't know but now I realise what was going on. Some of them were so pathetic, those young girls.

'There were traffic lights at the junction. I will never forget on one occasion there was a couple of old ladies came along in an Austin Seven and they were travelling from the north to the south and they crossed the lights and unfortunately when they crossed them they were red and they got to the middle and realised their problem and they hit the kerb on the other side of the road. Went over the dip into a privet hedge which was surrounding the Six Bells restaurant. They went crashing through there and they ended up upside down on the Six Bells' ground and they came out completely unscathed.

'1942, I started working there. It was Bignall and Cutbush Nursery. I would be sixteen, I suppose. I was just an ordinary working man on the nursery. Just a hand. There was a Royal Agricultural Committee. They were in charge of what we grew and what we didn't grow, so it was concerned a lot at that time with food production. We grew potatoes, cabbages, peas, beans, tomatoes, and lettuce, all common or garden things that you grew in English gardens. We grew on a fairly massive scale but we had these two or three Land Army girls helping us and we were a motley crew, I'll tell you!

'I was originally employed on what was called the Shrub Department and after a while I transferred to the Herbaceous and Rock Garden Department because I wanted to specialise. I didn't want to be just an ordinary chap on the nursery, I wanted to get to know something about propagation for instance, so I transferred and I worked in the glasshouses.

'To be honest there wasn't a great structure. They recognised the fact that you could do this job or that job so they gave you the job – you didn't get extra money for doing it. The only people who got the money were the foreman, the under-foreman and the foreman of the Herbaceous Department. There was a chap called Gillson and he was the chap I went to work with eventually in sales. It was a better department, the people that worked there were my sort of people. What I am trying to say is that the old country people, they were lovely. Those old men, they really were, but they were just old country men who hadn't got much clue about anything apart from what they were doing and the way they lived, the way they wanted to live which was fair enough. But I was interested in, as a young man, poetry, and book reading, and learning and the people I met in the Herbaceous Department were of that ilk. They had been sent there by the Government because they were, I think, conscientious objectors and one or two had been to university and one actually had a bookshop in

Park Road near Baker Street. He was a nice chap; he used to read *The Times*' "Literary Supplement"! Completely out of place in a nursery you would have thought but that was the way he was and I liked those sort of people, I got on with them, I learnt from them.

'There was an old shed that we used, a big old rambling shed that was the so-called office and an old-fashioned stand up and beg telephone and Mr Wickens was the foreman – Fred Wickens. He lived in Black Horse Lane in South Mimms, and his assistant, the under-foreman, was Charlie Stokes. He didn't make old bones like Fred Wickens.

'I used to cycle there. I lived in the village (Borehamwood) with my parents in those days. I used to get up at quarter to six and cycle. I used to have to be in work in the summertime at seven and in the winter half past seven. I cycled up the A1, the Barnet Bypass as it was then. I used to pedal like mad trying to get to work; I was worried about the time because the foreman, Mr Wickens, used to stand on the front by his office with his pocket watch out looking for you.

'We used to pop over to the Middlesex Arms sometimes in the summer; in the lunch break I would go and have half a mild which was quite a drink in those days. They have pulled that down, of course, some time ago before they did the Welcome Break area. I can't think why because it was a lovely old pub. Well, it wasn't that old, it was built in the old style. I think it was built in the 1930s. Perhaps the trade wasn't much. You don't recognise it now, it's completely different.

'I have been to the Welcome Break service station. When it first opened I went. Living here you don't need to go there because you are so near home.

'It's certainly different now! In today's world with the volume of traffic it's inevitable that it had to change. It's a good thing for travellers; it's a pity that the countryside had to be spoiled which, of course, it was. People who travel do need these facilities. Toilets are jolly useful when travelling, and a bite to eat and a cup of coffee or whatever. But it's so big, so vast, I just couldn't believe that that would materialise from the old Bignall and Cutbush Nursery where I used to work.'

12 November

Chris must have got up the wrong side of the bed this morning. He fails to give me a 'free' chocolate with my coffee and croissant.

The place is deserted, bar a few lone travellers and businessmen. A woman is sipping fresh orange juice as she deliberately shifts paperwork full of coloured pre-charts. I am reminded of a Church of England Sunday morning service with only a few of the regulars and a couple of passing strangers.

The white 'Rastafarian' has appeared again with his metre-long dreadlocks trailing down his back. His multi-coloured climbing trousers match his bobble hat, although his white, kicked-about Adidas trainers look peculiar on his small feet. He smokes the statutory roll-up cigarette.

The Welcome Break TV is screaming in my right ear.

'Stay watching for traffic updates!'

'Harry Potter tee shirts, buy one and get one free!'

So far I've managed not having anything to do with this 'cultural phenomenon'. I must be one of the few people remaining in the western world (the whole world?) who hasn't read any of the books yet.

I check my lottery tickets during an interlude between watching an African couple and their young daughter, all dressed in leopard skin patterned clothes and a man playing with his personal organiser.

13 November

There's an icy wind zipping around the car park, but inside there is a lunchtime buzz about the place. Joan, a middle-aged woman, serves me my coffee.

'Sorry, we don't do decaffeinated.'

A mixed bunch of white men, all shapes and sizes, but look like the kind you stand behind at last orders in the pub, sit together quietly. One of them with a shaven head is staring intensely at me. I drop my head and concentrate on reading my newspaper.

There seem to be some incredibly tall men around, all carrying Burger King takeouts. Why? At 5'8" myself, have I missed out on something? One blue-suited male, white hair, with the face of a twenty-year-old has legs taller than me; he draws stares as he lopes out of the building like a giraffe.

An army troop marches out mob handed. The last one is a woman although I am not sure.

Two women are having a natter. As a man it seems to me that women talk differently than men. For one thing they nod and agree far more. It is friendlier and they maintain eye contact longer. Take three men facing me. Truck drivers, brown boots and all, tucking in to big lunch specials with pots of tea, they talk to each other as they dig around their food with their knives and forks. Words are interspersed with forkfuls of food being eaten, eye contact is minimal. They smile into their lunches and stare around, they all talk at once, and they all eat a mouthful of food in harmony. Silence together. It feels like watching synchronised swimming. Food eaten, it is sit-back time, face to face contact. Teeth are picked openly with an array of toothpicks manufactured out of cigarette packets.

14 November

Come out of the shop, narrowly having avoided buying a multi-coloured paint gun kit, into direct sunlight full in my eyes. Feel like I am on the set of *Close Encounters* when the 'missing' people come down the ramp and behind them is a bright shining light. Dazzled, I bump into a middle-aged man wearing a black anorak, bit of a serious trainspotter look about him. Is it snowing outside? His coat collar has a liberal sprinkling of dandruff over it.

The rural bourgeoisie are in. Standing next to me in the queue for breakfast is a man straight out of the pages of *The Field* magazine. He's dressed in plus fours (I didn't know they still existed), checked shirt, tweed jacket, and a yellow woollen tie. He stands in a pair of expensive Church's shoes. With a plummy public school accent he orders the traditional English breakfast. There are sniggers from behind the counter. Darwin, his friend, comes over. Far be it for me to stereotype people but he is another country clone, woollen suit, same posh accent, tall, gentleman farmer feel to him. He goes for the same breakfast plus double helpings of eggs and bacon. The meal is paid for with a Coutts Bank platinum card.

I notice how posh country gentlemen have a habit of taking

over, of not seeing anyone else around but themselves. I am pushed to one side as one of the twosome leans over to collect his pot of coffee – no 'excuse me'. They have tied up three staff with their order whilst the rest of us peasants stand and wait. The staff are jumping around to serve them. Good to see the class system in England is alive and kicking, although the latter is perhaps what they need. Now I know a couple of mates in East London who could do the business!

We eat our meals to the sound of Spanish guitar music emanating from the TV screens above our heads.

15 November

Is there a business convention on? Lots of suited middle-aged men and one younger blond businesswoman with a red poppy in her buttonhole stand and sit around. A very large American male walks amongst them, his 50-plus inch waistline is hidden beneath a tent size jacket. His corporate tie and shirt give him an impressive appearance.

'Hi, everyone.'

Mobile phones call their owners as they talk and exchange pleasantries. A few hangovers linger over the breakfast gathering. The American 'Mr Big' sits across a chair and a half. A Sheriff of Nottingham character in a red mountain jacket ticks names off a list. Three people are lost behind Mr Big's Atlantic Ocean size shoulders and so are not ticked off.

They're off, this motley collection of business movers and shakers. Mr Big leads them out with the Stars and Stripes flag on a pole; it's the 1984 Olympics all over again.

16 November

It's Friday lunchtime and the place is heaving.

Wallace Arnold coaches are stacked up outside. Inside 'hobbly' old men and overweight women scoff tea and doughnuts. It is difficult to get a seat amongst them and the baby feeders, business meetings, secret liaisons, weary motorway travellers and assorted oddballs.

I sit alone. Which one am I?

A man in a hospital consultant 'uniform' of light-blue button-down shirt, plain tie and cream coloured trousers plays the Homer's Meltdown machine whilst he munches on a sandwich.

Three young men with women on their backs charge through the main concourse. They are joined by fifteen more, and an older man. He calls them all together and they group together in a circle, hands linked, then down on their knees, shouting and chanting. A Gregorian chant? Lots of whoops and 'here we go again'. They line up at Burger King for their 'nose bags'. They are students from one of the local colleges. It's Rag Week. The Portuguese table clearers are not convinced and call for assistance as they chatter amongst themselves.

Two ageing long-haired bikers, the types you tiptoe past on lonely roads in the States, are chaperoning a couple of young adults with profound learning disabilities. The longer haired biker with a ponytail and big boots has been over to the 'leader' of the rag week group, who is wearing a joker's hat with bells on. They face up to each other and talk as the two young adults make whooping noises. They are led out in a daisy chain by the bikers. The rag week celebrations start up again as soon as they exit.

*

'Some of the most important exchanges go on in the toilets.' (Professor William Brown, ACAS, on the negotiation between the firefighters and their employers, Observer *8 December 2002)*

*

17 November

Gents' toilet; a mobile phone rings.
    'Hello, Tel.'
    'South Mimms.'
    'Where did Nobby go?'
    'He's a squeak, didn't even see his fucking dad.'
    'Who came in right at the death?'
    'Lager's tasty.'
    'Nigel rang me, what time did I get home? Fuck knows.'
    'Bring the bike round to you, Tel.'
    'Did you?'

'Got you.'

'Gave him a score. Wouldn't knock it off the bill. She had got a lot to say for herself. He was talking and eyeing up other birds. She's standing there. Yeh, it's fucking demeaning.'

'Do what, Tel?'

'They won yesterday, Tel.'

'Where you off to today, Tel?'

'Off for a drink with Jason?'

'Yeh.'

'Yeh.'

'Got yer.'

'He rang me up. Said what time did I get home.'

'Fuck knows.'

'I had the hump over that. The cheque came through from the council. Phoned the geezer up. His name was on the fucking cheque. Printed out, Tel. The scaffolders have got to be paid, Tel. They want it for Christmas.'

'Fucking cunts.'

'Phone Eddie up.'

'Well, that's it. Do you know what I mean, Tel.'

'He's not putting himself out, is he?'

'Cheque went to the builders. No fucking good. We used company headed paper. Some prick in the council got confused. Said I'd take the fucking tiles off the roof if he didn't pay. They're my property.'

A racking smoker's cough reverberates around the toilet.

Laughter.

'He never showed did he?'

'Well, that's it.'

'Yeh.'

'Yeh.'

'Oh, it is.'

'Got yer, Tel.'

'Yeh.'

'Cunt likes expenses. No, that's it, fucking hell. Good company, Nobby.'

'What? That's the little mob down the other one.'

'She's all right and the kid. Think you're right about her.'

'That Nigel's got some spiel.'
'Did, didn't she?'
Silence.
'All right, Tel, I'll pop that round.'
'Chelsea and Liverpool playing today.'
'All right, Tel, better get back to madam.'
'All right, Tel. See you later.'
Silence.
Mobile phone rings.
'That you?'
'Yeh.'
'I am in the toilet still.'
'Don't know what time I got in 'ere.'
'I'm on me way, babe.'

19 November

Why is it some couples wear matching 'syrups of figs' (Wigs to the non-East Enders out there)? There I am dozing over my coffee and paper, listening to the strains of Blur, and lo and behold a middle-aged couple sit down, both with artificial hair. He is in his mid-forties, with dark rings under his eyes, which are hidden by Nazi German style glasses. Dressed casually, his hair, or rather his wig, is an off-brown colour which reminds me of old brown shoes which have been repeatedly brushed and polished. His wife is wrapped up in a thick grey coat and black trousers. Interestingly she has red rambler's socks on tucked into her black buckled shoes. Her blond curly wig moves up and down and she sips two-handed from her tea cup. She looks older than he does. Amazingly he has an Alistair Cooke voice, complete with long pauses. I am mesmerised by the wigs and his voice. In the background a Cher song seeps out of the overhead TV screen.

A white porky pie man sits down with his head resting on his hands, and his elbows on the table, his eyes lost on the Asian woman facing him. She plays with her knees and moves her head around. Another hospital dressed consultant dashes past on his way to theatre. His hands are now together in prayer fashion. She's interested but not that interested.

I manage to squeeze another cup of tea out of the pot. With these prices you tend to.

'Finished?'

Unknown to me a table clearer has been watching me and is now alongside, her hands reaching out to take my tray and assorted rubbish.

She smiles and disappears with my tray. Now this is unheard of, a table being cleared whilst a punter is still at it! I feel honoured. Get the feeling the table cleaner staff are simply acknowledging that this man, i.e. me, sits at tables longer than most other punters and is always writing. I feel vulnerable. Do they stand and watch me, and talk about me?

The white man and the Asian woman are joined by a Teddy Sheringham lookalike, but with brains. He gesticulates rather too much in talking to them and begins to look like a half-pissed tic tac man.

21 November

A long queue at The Granary today; they seem to be short of staff. Two empty trays stand on the counter.

'Is Security coming?' asks a harassed cashier.

'That's £4.57 love.'

'Where's the butter?' I ask.

'On the counter.'

I take a bagful.

A large group of blue overalled young white males with blue berets and black boots stand around outside the Game Zone. They have name-tags sewn onto their breast pockets.

Feel a bit queasy watching Mr and Mrs Fat Northerner eat what looks like double breakfasts.

There's a film crew about, large white lights, microphones wrapped in best Wilton carpet. Young pretty things with Home Counties accents mingle with the 'roughs' who move the equipment around. Some people stare, hoping to catch sight of an EastEnders TV soap star.

22 November

It is raining outside. Bit of a hangover this morning – too much Draught Guinness last night with some Australian friends down in Brick Lane.

The lime pickle from the Indian Restaurant we ate in forces me into an extended stay in number 15 cubicle. This causes some concern to the toilet cleaner whose regular rota of cleaning up after us is brought to a standstill by my predicament.

A tap on the door.

'Hello, are you all right, mister?'

'Eh, yes, thanks.'

I avoid his concerned look as I quickly wash my hands and head off for a coffee.

A 1970s man appears wearing a brown leather jacket with 4-inch wide lapels, and a multi-coloured hooped sweater. I am sure he's wearing flared trousers as well.

He smokes a large pipe which I haven't seen since the days of standing on the terraces at White Hart Lane cheering on Spurs with my late father. His companions are more of this century although they have that 'best clothes' look about them. They are going somewhere.

24 November

What am I doing here?

Cold draught of air blowing down on my neck from an overhead air conditioning system. Feels as if I am sitting drinking coffee in a refrigerator. Still, not all's bad with the world. Arsenal were beaten in the Champion's League football match the other night and last night's Channel 4 programme, *Being Mick*, a self-penned portrait of Mick Jagger by himself, gave me a few laughs. Apart from The Who guitarist, Pete Townsend, giving a believable impression of living on another planet, it was good TV.

Pity then about this morning's banal drivel from Welcome Break's satellite TV station, perched, parrot-like, on ledges above our heads – a mixture of blank screens and 'music' by artists I have never heard of.

What is it about couples, men and women, walking arms around each other's waists so early in the morning? Have they just emerged from the Days Inn Motel opposite where a night of pleasure has occurred? Or is it simply they are sad people? I suppose it could be love the way they stare into each other's eyes whilst walking. It's like watching a couple of thirteen-year-old adolescents' first love.

Couple of Muslim women dressed in veils turn a few heads as they slowly parade through the main public area en route to the toilets.

26 November

Groups of cheery little elderly people, men and women arm in arm walking around, flat caps and sleeveless tops, permed hair, chewing mints in tandem.

Two British Airways stewardesses in uniforms sip coffee secretly and look out for male admirers. The Greek businessman sees them as he talks loudly in to his mobile phone. The remains of his Danish pastry are cleared from his front teeth with the corner of a cigarette packet. The 'saved' pieces of food are recycled.

Lots of people wearing sunglasses. Well, it is late November and the sun's shining.

Some pensioners, who look as if they have lost directions to the nearest bowling green, walk aimlessly around.

'It's out of the door, turn right, and about 25 miles down the road.'

'Thank you, young man.'

A confession. I've stopped wearing my fleece when I come in here now. Reason? Well, with my greying hair, okay, grey hair, I am beginning to feel stereotyped by some staff and other customers. You know, that smile younger people give you which means, 'God, I know I am serving you and you are a respected and valued customer, but please, please hurry up and make your purchase and piss off back to your coach with the others.'

27 November

A short, chubby man with a ginger ponytail and goatee beard plays the machines in the Game Zone. He is by himself. His left hand is in his trouser pocket. It's 7.50 a.m. on a cold morning.

'You playing?' he asks me.

Difficult to tell where the accent comes from. It moves between the North London postcodes like an illegal mini-cabbie.

We take turns in shooting up the raw recruits as they appear on screen. A sign informs us 'not to play this game if you have been drinking'.

29 November

Major accident on the motorway this morning. Seems like a horse box has decided to go walkabout across the central reservation. No sign of the horse as traffic creeps past the aftermath. We all drive slow and courteously, coming away from the scene of an accident which has completely closed the eastbound carriageway. This soon gives way to roaring traffic, focused stares and the normality of the 'Fuck you pal, I am late for my meeting now' mentality. I am lost in the spray as we speed over the brow of the next hill.

The tranquillity of South Mimms is reminiscent of a Scarborough tea room – the gentle hum of conversation, the chink of cups, papers rustling. I look around for the pianist. Outside someone (a horse?), may be dead or seriously injured. Here I am cocooned in the M25's inner sanctuary.

30 November

It may have been the effect of too much caffeine but I am certain I overheard someone singing 'Nessun Dorma' loudly in cubicle 6.

*

Motorway service areas exist to fulfil a road safety function by offering motorists an opportunity to stop and rest. (House of Commons, Hansard, April 3, 2001)

*

1 December

Passing through in between journeys. It's belting down outside and deathly dark.

KFC has a long queue, mostly middle-aged people. Burger King next door is touting for customers with special deals. It's attracting the younger end of the market.

A Dot Cotton from EastEnders patrols the aisles with her tray of tea. She looks and gives me a wide berth. Her tea cosy hat is tilted to one side like an old Thames barge listing to port.

A family of eight people in front of me are consuming a variety of food from different outlets. One of the two dads' moustaches twitches as he mauls a burger. His fellow dad has sunken eyes and speaks in short bursts; their daughters look on. He has a Jonathan King smile.

I carefully cut up my almond slice and catch the woman alongside me eating an oversize burger with both hands. She has red dreadlocks spilling over a pair of broad shoulders. Her companion talks whilst he eats; food is sprayed around. Her napkin drops to the floor leaving her coat sleeve to wipe her mouth. I take another slice of almond. Jeff is looking for a machine to play on. 'It's over there, Jeff.' Thick northern accent. Her sleeve wipes tomato sauce off her chin. I notice that as Jeff seeks out the Game Zone his shirt tail is hanging out; it is dirty and stained with brown streaks.

The Burger King staff on duty all look fifteen years of age; the girls in green baseball caps and the boys in blue caps. Surprising the number of people who leave half-eaten burgers and fries lying around the tables. Mind you, the guy standing at No. 2 'feeding hole' looks as though he could clear all the burgers awaiting customers.

'Is this seat taken?'

A group from a coach sits next to me.

'I'll sit here with you to help you out if you can't eat it all.'

She has a jam doughnut, fruit and cream, and a plate of chips.

'We've got twenty minutes before we leave, eat up.'

'Does the driver know where he is going?'

The table cleaners are working their socks off. J-cloths whizz over the tables.

The coach driver with manicured short hair sits alone, a large Coke at hand, as he tucks into a late big breakfast. Naval tattoos of naked blondes on his arms dance up and down as he eats.

More coachloads of senior citizens are arriving, spilling people into the bright lights.

'Reminds me of Blackpool, Eve.'

'My Tommy would have liked this.'

They wander around like children at Disneyland, looking but not sure which 'ride' to chance.

Catch sight of a man wearing tartan carpet slippers on his feet. He is on a mobile phone. He's joined by a man wearing a sleeveless fishing jacket with many pockets. They both smoke roll-up cigarettes – it's a 'no smoking' area! They seem mesmerised by the dancing, singing women on the TV screen above them.

A man in a London Underground fluorescent jacket, clutching a mobile phone, walks in and looks around.

'How long to wait for the Victoria Line Brixton train, mate?'

2 December

Bit quiet this morning, church-like atmosphere, a hushed silence, bar the occasional giggle from a group of male Verve lookalikes. It's feeling a bit like my front room at home. Kind of comfortable, familiar; where's the cat? The TV in the corner or rather over my head, the chairs now have a welcoming shape about them, my body easily rests into them, table at the right height, zip-up slippers and pipe, M and S cardigan. I feel as if I am in my late father's homemade green house in Walthamstow. Along with the tomatoes I am being nicely warmed under the arch lamps that hang from the roof.

8: Trouble with the Saga Louts

> Around such common places lives are configured, remembered, and anticipated.
>
> *John A Jakle and Keith A Sculle*

A Christmas tree from Norway? – give the Ultimate breakfast a miss – promenading – Jay's visits – football supporters are doing my head in – quack, quack – Christmas Eve – bother with the 'oldies' – Truckstop – get my photograph taken – spoilt middle-class kids – South London family – New Year normality.

3 December

The builders have come in out of the fog; skinhead style haircuts, green combat jackets popular with members of the far right British National Party, splattered in white paint, and trousers in need of repair, particularly around the crutch areas.

The seasonal Christmas tree has arrived. From Norway? A suited man looks at it in wonder. Others walk past without noticing it.

Two more builders appear to meet up with their mates. Familiar brown steel-toe capped boots, hands in pockets and big bellies crying out for Fat Boy breakfasts. Why is it builders always have paint blobs on their trousers? Is it a mark of being a 'real' painter? Like a sign which tells the surrounding world, 'Hey, I've been there, hell it was tough… some paint got spilt but we kicked arse.' Or are they just cowboys?

4 December

Feeling a bit peckish but decide against having the Ultimate Breakfast and go for a pedestrian two eggs on two toast. The queue is a mile long. New staff on this morning.

Parading up and down is a 40-something-year-old woman wearing a short black tight skirt.

'She's showing the tops of her black stockings,' my blond-haired woman neighbour in the queue whispers in my ear.

'Not bad,' a blue-suited businessman next in line to her remarks loudly to his colleague.

Maybe it's the see-through black top that she is wearing which collectively turns the heads of the males in the breakfast queue. She returns, her slip-on backless shoes clipping over the tiled surface.

'It was two eggs, sir?'

'Yes,' I reply, my attention still drawn to the promenading in front of me. She sits down at a couple of tables up from me.

It's taken fifteen minutes to get my order in but the waiting was worth it. God, how I love brown sauce.

The man in front of me, a chubby fresh-faced thirty-year-old but looking fifteen, can hardly believe his eyes or is it luck? Legs open, with a full view of her 'wares', he shuffles in his seat as if adjusting his 'lower carriage'.

I am not sure whether to laugh or cry as she fiddles with her silver mobile phone. Her elbows rest on the *Sun* newspaper. Strangely she reminds me of Mary Quant who I once saw in Biba along the Kensington High Street in the 1960s when I was waiting for a girlfriend who was looking at some bras.

I keep my head down as her eyes scan the horizon. I notice I am not alone in doing this. She takes her coffee and paper off for a cigarette.

The normality of the morning businessmen's club returns.

Bump into a 1960s time warp man complete with Afro haircut, dark glasses, high-heel boots with blue Levi's jeans hugging his legs. His wife looks a bit like a northern housewife, long hair which would suit a younger woman, and mail order clothes. An unlikely pair, they seem lost. I feel tempted to direct them to the 'sad people's' enclosure.

People seem to be entering my space this morning.

A businesswoman with freshly ironed straight hair sits down in front of me although there is a sea of empty seats around. She is in a serious black business suit with matching shoes, an important looking brown leather Filofax (large size), with matching small black handbag. A copy of today's *Times* newspaper sits with her. She sips her coffee as if she was drinking a martini. Why has she badly chewed fingernails?

Two burly policemen call in for their breakfast, their Sig Sauer P226 9mm semi-automatic hand-guns neatly tucked into waist holsters.

Passed the Tourette's man in the car park. He looks at me intensely. I feel myself nodding at him. We pass each other, shoulders almost touching.

5 December

Bad hangover. An early office Christmas bash yesterday has left me permanently scarred. Seem to remember ordering a coffee and getting lost in the gents' but not much else!

7 December

Jay is a tall woman in her mid-twenties with blond hair. She works at the nearby university.

'Well, it's not really a stopping off place on a journey for me because it's so close to home; by the time you get here you could be home. The last time I was here I would have been eighteen or nineteen, so seven years ago now.

'When we used to come here there would be a group of us out for a night and it was just somewhere to go. Just a kind of winding down sort of place; just get a cup of tea – that was the idea – go somewhere for a cup of tea. Kind of strange, but we just used to sit and have a bacon sandwich or whatever and a cup of tea and have a chat.

'We used to do it now and again after we had left the pub. If the evening had been too short sort of thing, which would be about half eleven at night, we used to come from a club in Edmonton, North London, which used to chuck out at six in the morning.

'There was always me and a couple of girls whom I was friends with. Usually a group of half a dozen blokes I used to hang around with, so any combination of those.

'You would see a lot of boy racer types so it was a chance to see them showing off their cars. That was all part of the fun you see. You would get little Escorts and little Fiestas, seeing who could drive the fastest. It was like an initiation sort of thing.

'It was, I suppose it still is, a kind of strange place because it was never shut and you used to get coachloads of German school kids at three in the morning, and other people wandering in. There always seemed to be kids playing on the fruit machines and video games and I thought it a bit strange, like what are they doing there. Fourteen- or fifteen-year-olds, that sort of age.'

10 December

It's cold and foggy outside. Inside the lights shine brightly and people go about their Monday morning business. Martin Creed has won the £20k Turner Prize for his 'The Lights Going On and Off'. I get the same effect after several pints of Stella Artois and listening to Pink Floyd's 'Great Gig in the Sky' track off their *Dark Side of the Moon* album.

A large group of older Muslim males enter, all dressed in white robes. They have young enthusiastic 'minders' either side. Heads turn as they proceed to the toilets in a conga line. A few sniggers from a group of white and black men sitting together.

12 December

Skip the photogenic Ultimate £6.99 Breakfast despite its seeming popularity this morning. What is it about this M25 version of the traditional English breakfast that people of different backgrounds and professions eat it? Some young lads, university students, and a couple of budding film stars except it's a mother and her daughter, difficult to make out which one is Joan Collins, are all going for it big time! Me, I go for my usual safe two eggs on two toast and a pot of coffee. They're at Geneva prices, of course, but what the hell, it's Wednesday; I am halfway through the week and feeling good.

I buy all the morning's papers to review what's being said about the reports on the 'race' riots in northern towns this summer. The cashier greets me with a 'Good morning' as I plonk them down on his till, but eyes me with suspicion.

15 December

It's a sunny Saturday morning. The two Leeds United footballers, Lee Bowyer and Jonathan Woodgate, have walked free from Hull Crown Court yesterday for an attack on an Asian student. Why?

I am in the queue with my *Guardian* newspaper and a copy of the *Sun* (for research purposes of course!) waiting to be served by Remi, a mixed-race woman. Standing next to me in the queue is a beer smelling football supporter buying his *Sun* newspaper and a Mars Bar. Across from us his mates enjoy some humorous banter with him, which is tinged with comments like, 'She'll do you alright', 'Have you washed your hands?' Just good friendly racist jibes. I feel the need to put the boot in, particularly as they're Arsenal supporters as well. What would Nick Hornby do?

The place is heaving with travelling football supporters complete with Star Wars tuned mobile phones. Much as I like the films, too many mobiles with the same 'Empire Strikes Back' theme can do one's head in. What is it about football supporters? That 'thick' look, the *Sun* and *Daily Star* papers tucked under their arm as they adjust their tackle hidden somewhere under expensive designer jeans, their women, only a few of them but noticeable, with their 'blokes look'. Types you would not want to bump into on a dark night. One green-shirted supporter, an away shirt so I have no idea which team, has tattoos up his arms like an army regular. He eats a double breakfast, sensible in case the match goes into extra time I suppose. But I am fascinated by the tattoo on his neck – it's a zip. Maybe his tackle is up there?

'Excuse me, mate, is this a smoking area?'

'No, over there.'

Read the fucking sign, bonehead. They have failed to read the 'Thank you for not smoking in this area' signs on every table.

We are joined by a group of muzzled young players offloaded from a Nationwide League football club minibus. They roam the concourse like a pack of hungry hounds looking for a fox. Got to give it to them though, smart blue tracksuits with white trainers. The group enters the shop and reappears with their daily fun papers and more sweets than my daughter used to crave for when she was three years old. They stand around and look at all of us, we all stare back. Eerie!

Three large 'minders' follow in their wake. They remind me of ex-boxers who have had one fight too many for a last pay day, 'money for the missus and the kids', who end up being bouncers on the doors of East London boozers. One is on his mobile, walking around with it, talking loudly, gesticulating wildly and throwing his body around with the same disregard a builder shows for other road users when driving his firm's dirty white transit van. He clutches a bottle of Lucozade, somehow managing to scratch his crotch at the same time.

Two green-blazered coach drivers tucking into hot meals look at him impassively.

Later...

Happy families are experiencing Burger King parties, whilst untethered fourteen-year-olds run around. A young lad, the same age, chases after them in his motorised wheelchair; his mother watches from a distance.

Mr 'Potters Bar' man out with his wife for an afternoon tea look on and start making duck sounding 'quack quack' noises.

18 December

It's cold inside this morning.

There's a distinct lack of businessmen and businesswomen around this morning. The place is full of 'drifters' and other people passing through. Seems to be a glut of middle-aged men wearing an assortment of baseball caps. One has that leather version you sometimes see 'oddballs' aka BNP paper sellers wearing at the top of Brick Lane, East London, on a Sunday morning.

Watch a group of people try and make pleasant conversation with each other. They are working for the same company but are strangers. They sit a safe distance away from each other, a mixed bunch. Two African men, a sprinkling of white men and women, and one Muslim woman in a veil. A dynamic blond woman, fresh off a corporate assertive team leaders' course, greets the group and leads them out.

24 December

Christmas Eve. The coach park is full. Coaches awaiting to depart to the corners of England and Wales, Chester, Lake District, Torquay, Vale of Glamorgan, Pennines and so on. The car park has reached gridlock. Inside I am confronted by hordes of elderly people. There is nowhere to sit. New grey permed heads bob up and down, flat caps and bald heads carry trays full of teas. Bags with Luton Airport labels bump into people's knees.

An air of urgency pervades. There are a couple of women clutching their trays. They look over the seething mass for their loved ones. I expect a call to go out, 'Will the parents of Maude and Millie please collect them at the La Brioche Dorée entrance.' I see them waddle off into the distance with their trays.

The Welcome Break TV screens are silent, or is it I can't hear them over the chatter and noise?

The 'table clearers' are fighting a losing battle. Disgruntled senior citizens are clearing tables for them.

A man in a long black raincoat with shoulder pouches (for his cap) sits across from me. He is a ringer for my Uncle Les, who swore the same German Panzer tank chased him on his motorbike across France to the Dunkirk beaches. I can still picture him at family Christmas parties, drunk, but happy, reminiscing about this experience.

With his brilliantined hair, and sunken cheeks, which are reddened by working outside, his nicotine-stained teeth chomp into a jam doughnut. His wife is covered up in a scarf which reminds me of my mother's red and white check tablecloth she saved for Sundays. Both have hair growing out of their ears.

One group of greyies have a fifty-something woman with learning disabilities sitting with them. She is dressed like her mother and the couple across the table from them. They are in their seventies. She blows bubbles as she listens intently to the table conversation. She's there but no one talks to her.

After centuries of marriage couples sit and eat and sip their teas in silence, faces staring into space, conversation is muted.

The fight for seats continues. Some people are queuing for a seat at a table. What is it with the British that we form orderly queues, whereas in other parts of the world it is first come first served?

A lone woman, in a maroon M and S pensioner class jacket, neatly brushed hair, blue trousers with knife edge creases, eats her cake alone. Her wedding ring hangs limply from her finger.

It's time to go. Slowly they rise from their chairs and walk deliberately to the exit for coach passengers. The noise level drops, although the bugle being blown outside temporarily stops this. A plate crashes to the ground from where the crockery is being 'processed' by the table clearers.

Does it feel like Christmas Day tomorrow? Apart from some staff, a few people wearing Christmas hats and a Christmas tree lighted up, I am not sure.

More crashing of crockery. A pensioner in an Afghan hat puts some red lipstick on, stands up and smiles at nobody in particular. Her false teeth slip as she follows the crowd.

27 December

I have to queue on the slip road coming off the M25. I queue to get into the car park and I queue for a car space. I queue at the gents'.

The coach park is alive with pensioners and their suitcases. I feel like the last 'young' person left on the planet as I weave my way in and out of a tide of determined grannies and granddads rushing to and from the toilets to the tea and coffee outlets.

A thirty-handed mob of 'Saga Louts' come in from a 'Chester for Christmas' coach. They rush the orderly queue I am standing in waiting for my coffee.

'Do you do hot chocolate?'

'I want two... no, sorry, three.'

Queue hoppers are everywhere!

A man walks the length of the queue dressed in a belted green mac carrying a small brown case, with his name and address emblazoned on a label on the side of the case.

Ian, a man in his late sixties, puts aside his Rebecca Shaw novel, *Trouble in the Village,* to take his false teeth out of a black plastic bag. I watch, mouth open in horror and amazement, as he proceeds to fit the upper and lower denture set into his mouth in preparation for his tea and cake which his wife has dutifully brought to him. He has that faraway look as he munches into his Danish pastry.

A couple of teenage kids, both with telltale signs of having a learning disability, trail their elderly parents like puppies.

I think I might have to have a complete re-think about dyeing my hair. It's currently what I like to call a 'sophisticated grey'. A couple of senior citizens, female, are giving me a particularly 'unsophisticated' come on sign!

I sit with Wayne (a dead ringer for Sting) and Matt. They're both off-duty firemen. Both are working part time here loading suitcases on and off the coaches. I can't quite make out if it's cash in hand they get or whether they just do it for the laughs.

It's getting dark outside. Traffic is still moving slowly out of the car park exit. Coach passengers are gathering their bags. An announcement calling people to board is lost amidst the steady hum of tea and coffee chatter.

Decide to go walkabout. Wander over the darkened tarmac of no man's land to the Truckstop. A sell-everything shop, including condoms, 'fronts' the lorry drivers' café. As I walk in pushing open the door, I'm sure the music stops. I'm confronted with men sitting at individual tables facing me. Smoking is, of course, permitted here.

Nervously I stumble over to the counter and ask for a coffee. The disinterested Nigerian member of staff points to the large polished silver urn on the counter which is the help-yourself coffee machine. Nowhere to put the money in. No instructions. No one comes to my assistance. I grab a stained white mug and stick it under a white plastic spout which is protruding out of the mother urn. I manage to overfill the mug. Coffee floods the tray. Milk is despatched from a white jug which requires a twisting of the cap and lining up two parts before milk emerges in stops and starts. I succeed at the third attempt.

'Easier milking a fucking cow, mate,' a Geordie trucker with boots the size of skis whispers behind me.

I grab the nearest empty table carefully avoiding eye contact and stare at the football pitch size television screen.

The muddy fluid in my mug cost 95p compared to £1.79 over at the service station.

Jill brings out meals from the kitchen.

'Double Large Burger, ham with extra chips?'

'Over here, love.'

'Number four breakfast?'

A plate the size of the county of Surrey is carried head high over to a table behind me.

'Bacon and egg sandwich with steak and chips?'

An arm is raised.

For every order Jill brings out a large basket full of every conceivable sauce. Tabloid newspapers and dirty dishes litter the tables – no table clearers here.

'Cow horn burger with double fries, and four eggs?'

These guys are serious eaters. One of them sitting in the front row, bald headed, watches everyone coming in.

'Gammon.'

'What's that?'

'Gammon.'

'Go on, pet, I'll have it.'

A middle-aged woman comes in, stares at the fifty pairs of male eyes looking at her and runs out.

'Holding back the tears' by Simply Red is now playing above the sound of the television which seems to be showing a Norwegian soap which I'm not familiar with.

I become fascinated by the two sets of traffic lights hanging from the roof above my head. They are stuck at red! Below a nicely decorated family Christmas tree with a fairy on top hides in the corner alongside the giant size television screen. Alongside it stands a video screen complete with ghetto blasting speakers. Do they hold private parties here?

'You ham, love?' Jill asks me.

'No thanks. Actually I'm vegetarian.' Somehow I manage to swallow hard saying it so she doesn't quite catch what I said.

'What's that?'

'Just saying I've just eaten.'

She eyes me with suspicion and retreats to the kitchen. Moments later I see her peering at me through the service hatch window along with two other kitchen staff.

I smile back and retreat to the sanity of the service station across the car park.

Note: Discover later that the previous owners of the Truckstop had been fined for breaching health and safety regulations after a customer drank coffee from a pot that was cleaned with chemicals that had not been washed out afterwards. Spend the rest of the day drinking pints of milk!

*

'There are tailbacks from Junction 23 of the M25 at South Mimms Services to the A1M. Look out for that one.' (Radio 5 Live Traffic News)

*

28 December

Fifty-plus coaches in the coach park. The place is like an anthill. The gents' toilet smells of stewed tea, stale beer and damp farts.

Struggle to find a seat because of the crowd, parts of which feel like walking through the 'dead'. Manage to get a high stool after wrestling it away from an elderly woman.

I am sitting in the 'smokers' section' although surprisingly there are few smokers.

Someone takes a photograph pointing the camera in my direction – I smile – the flash causes some laughter, lots of gnashing of false teeth.

The table clearers are working overtime to keep pace with demand. I note a couple of Portuguese swear words as crockery crashes to the floor behind their screen.

Get a call on my mobile. Struggle to extract it from my black leather jacket pocket and in doing so miss the call.

'I hate this family!' a young boy shouts as his hand pulls out some fries from his Burger King box.

Decide I am right in taking the moral high ground by being a vegetarian as I face a family tucking into what looks like burgers of undercooked critter meat with red blood dripping down from them, although it could be tomato sauce.

The crowd is thinning out rapidly. The coach park exit has a slow moving mass of people moving towards it. White heads bob up and down. Normality returns. A member of staff brings out the carpet sweeper: a brush and pan to be precise. There are now empty tables and chairs. I feel vulnerable, like an oasis with desert all around me.

A large red bin is wheeled out of the main concourse and through the mysteriously marked 'Staff Only' doors. I have visions of an elderly person who has missed their coach, has been found on the floor, and is now inside the wheelie bin en route to the 'pensioners lost' section of the complex.

A 'mean' gorilla walks in with his two teenage daughters, balding at the front, hair slicked down at the sides and back over his eyes, wearing statutory expensive trainers. I can see how 'urban myths' arise. He is carrying half of Ratners jewellery shit on his fingers, wrists and neck. His two daughters follow him around like yapping baby seals at feeding time, both are aged thirteen or fourteen although they could pass for twenty-year-olds in a Friday night pub after a few beers. His daughters are queuing for KFCs whilst he sits on a stool scratching his armpits.

29 December

Plenty of lookalikes around. The late Uncle Albert out of *Only Fools and Horses*, is eating with his wife, and the ex-singer formerly known as Cat Stevens is there with his family.

Mr and Mrs Middle Class parents are with their two small children. Emma and Jack are being given numerous alternatives.

'Now do you want to go back to the car?'
'No, Daddy.'
'Would you like to look around the shop?'
'No.'
'Let's go and see what's over there.'
'Don't know.'
'How about the Game Zone area over there?'
Silence.

For fuck's sake just tell them rather than keep asking them! Poor kids, they don't really know what they want to do. So what happens? They start being bolshy with their parents. It's embarrassing to watch.

Sorry, but must mention that the red wheelie bin has emerged from the 'Staff Only' doors and is circulating the concourse – looking for other lost pensioners?

I retire to the gents' for a well earned pee. Carlos is happily jotting down his signature on the toilet wall chart which shows to the world that he has completed his half hourly cleaning of the toilet.

1 January

It's New Year's Day. The evening is cold, dark and frosty.

A woman is in with her narrow husband and two Rubik Cube shaped children. She has a 'raspy' South London working class accent.

"Ere excuse me.'

She has called over one of the table clearers.

'Can I get a receipt for my meal?'

No reply.

'I need it for my children.'

No reply. '…I get someone.'

She cannot wait. She marches off in her Adidas training bottoms and polished black shoes to the counter with one of the unfinished meals, her ponytail swings behind her. She has that determined look about her. A female table clearer has brought a male colleague over to them.

'She doesn't speak English!'

'I've brought you a receipt which shows the refund. I am sorry about that.'

The table clearer returns to her duties.

Two young people are dancing a slow foxtrot together in front of Burger King.

"Ere, mister.'

Who, me? I smile. She ignores me – she is talking to one of her children. She's lovely in a caring, motherly way, looking after all three of them. She keeps up a continuous chatter. Her gold hairband glistens. She delicately lowers more chips and pie into her mouth. Her husband holds his cup with two hands, quietly spoken; I can't hear him. The two children are aged 8-9 years, a boy and girl. The boy sits facing his father and the daughter facing her mother. I change seats to get nearer to them. The little girl walks over to my table, bends down and hands me a small carton of milk I have dropped on the floor.

'Thank you.'

She smiles and retreats to her seat. My eyes watch Dad looking at me. We both smile.

Travis and Lauren are two very well-behaved kids.

June has that working class South London aura about her; she sees all and is quick to decide if it's a 'yes' or 'no' – a hug or a slap. Every move the three make, she makes a comment. 'Leave it or you will get a smack.'

'It's cold, either eat it or leave it.'

She periodically coughs, hand to her mouth.

'That's it, Lauren, eat up.'

'Get on with your food.'

'Eat it, it won't hurt you.'

She has shiny white teeth like a hyena. Dad's voice barely filters through the rat-a-tat delivery of Mum. It's a cross between a slowed down version of Ken Livingstone's nasal drawl and one of my mates who used to work on a banana stall in Walthamstow High Street. He had a stammer so bad he had to speak very quietly and slowly so passing punters could understand him.

The little boy wants a headband like Mum. His question is returned multiplied by Mum.

'Your hair is lovely as it is.'

'Bands are for girls, stupid.'

She's up and off again to The Granary counter. Her daughter dutifully follows her. In the absence of both, Dad and son have a pulling match, the little boy trying desperately to follow them and Dad refusing him. The little boy swings his legs angrily under the table. 'Lauren, sit still.'

A jam doughnut and coffee have appeared. A packet of sugar is flapped vigorously around.

'Lauren.'

'Good girl.'

'Just one thank you.'

'Who wants hot chocolate?'

Both children put their hands up. Dad sneezes.

'I'll get a cab to the hospital tomorrow, Lauren. Come on, got everything?'

The table is piled high with plates full of chips and sausages uneaten, cups and saucers and Coca-Cola cartons.

I decide to go for a pee before setting off. I hear June's voice coming from behind the disabled toilet door.

'You wait there, Lauren, you hear me?'

'You wait there, Lauren.'

'Lauren, can you hear me?'

Dad is leaning against the photo machine clutching a Coke carton. The little boy is swinging around in front of him.

2 January

Get my brown Doc Martin shoes mopped by the anxious toilet cleaner as I sit in cubicle 7. I am suitably impressed by his willingness to clean the toilet regularly as duly advertised in the toilet entrance but I am not sure about my shoes as well!

A fifteen-year-old policeman is queuing at The Granary for his breakfast. He carries an impressive array of equipment on his belt as does his thirteen-year-old blond female colleague. Two equally 'young' colleagues join them. All four order their breakfasts. Another older (male) police officer has joined the four; he looks like Dad out for breakfast with his four young children.

*

You are watching Welcome Break TV 24 hours a day.

*

3 January

Normality has returned following the Saga Louts' invasion over the Christmas and New Year break. Businessmen and women are again conducting serious early morning meetings; laptop computers are sighted; families reappear with small children returning home in preparation for school next week; a couple groping each other emerge after a night spent in the motel; individuals are again passing through.

A man is moving slower than two Teutonic plates on a good day. Why is it that some (all?) very, very overweight people wear baggy tent-like clothes that hang off them, walk slowly, and deliberately never move out of the way as they walk towards you? A couple of innocent bystanders are squashed like hedgehogs.

5 January

Coming down the A1M from a business meeting in Peterborough, decide to call in. Confronted by two football supporters wearing tangerine-coloured shirts. Blackpool are playing Charlton in the 3rd round of the FA Cup. Both lads look a long way from home.

Watch a middle-aged man spend three minutes attempting to extract butter from a small plastic carton. He gives up and throws it over his left shoulder.

9: Bermuda Triangle

> The Bermuda Triangle is a mystery zone where thousands of men and hundreds of ships and planes have been disappearing for years without trace and utterly without explanation.
>
> Adi-Kent Thomas Jeffrey

Absent father – overstay my allotted toilet time – coffee with senior staff – works outing – brown-suited man – red is the colour – Dunn and Co – Mr and Mrs King – second-hand car dealers – I get interviewed – incident in the gents' – Alan and his 17,000 crossings – the three nuns.

7 January

The place is empty apart from a few scattered business people and a young couple necking who seem totally oblivious to people around them. The table clearers tiptoe discreetly past them.

It's one of those mornings when people who tend to watch game shows on TV, you know those with low IQs and short attention spans, wander aimlessly around.

8 January

Somehow the Tuesday morning dullness outside has penetrated the bright lights of the hangar. Whilst early morning table talk is of 'commitment', 'catching up', and 'performance', it's mixed with silent contemplation over bacon and eggs. Even the tinny rattle coming out of the TV in the clouds is politely ignored.

A couple of Eastern European-looking males, early thirties, negotiate the sugar packets for their coffees. One is blinking at the lights. The other talks incessantly. Both wear the same brown suede boots and regulation 'new in the UK' haircuts.

I get a smile from Tina, one of the table clearers. She cleans the table with mechanical efficiency. Her blue J-cloth misses nothing as it glides sensuously across the table.

A posse of heavily armed police arrive; their frames block out the early morning sun peeking through the glass exterior. A few heads turn. The two 'Kosovans' both stare in amazement at them. A businessman opens his large case and displays a plate with accompanying knife and fork strapped inside. A new type of laptop computer? The Kosovans leave quickly.

*

The Granary: A bright friendly self-service restaurant with a full range of hot and cold food.

*

9 January

Four joking northerners eating their KFCs. They wear uniforms of dirty blue overalls, black boots, and thick necks. They joke about an effeminate guy with ragged blond hair as he takes small steps on his long walk to the gents'.

A woman with a blank expression looks at them from afar. I see her eyes moving as she negotiates the tables. Has she had a nip and tuck?

Four South Koreans complete with airline bags tuck into full English breakfasts and large cappuccinos. They are all left-handed eaters and use forks only. Little conversation between the four.

'Hello, Tony.'
'I've got an agent with me.'
'His number?'
'GB321X.'
'Could you check his gaskets?'
'Quota?'
'Thanks.'
'DUBS49.'
'25341.'
'Tubs, yeh?'
'Top one, doing well.'
'What state?'

'These caps, have they got N power there?'

'Must go to the toilet; do you know where they are?'

Two green plastic Hovis bread trays, a long way from home, sit forlornly alongside them stuffed full of files and papers.

The Granary counter staff are standing in line waiting for the next supply of customers to mouth their orders. Like latter day fishermen waiting to get new arrivals from the bottom of the North Sea, they hold the necessary implements in their hands: potato scoop, serving gloves, large baked-bean spoon.

10 January

June and Roseanna are 'manning' The Granary this morning. I feel I am on a well-oiled conveyor belt as I am passed along with my tray of eggs on toast.

'Would you like your toast buttered, sir?'

'Would you like a drink, sir? Coffee or tea?'

'That will be £4.87, sir.'

A snip!

11 January

Two elderly Sikhs, a Mr and Mrs, sit facing each other sharing a plate of toast, beans and mushrooms, and a pot of tea with two cups and saucers. His white beard and turban match the colour of his socks. Both are dressed smartly. He is in a grey suit, she in a saffron sari, and glasses. Her hair is neatly controlled.

Two young women, white and mixed race, walk past and look at them en route to the still closed Burger King outlet. Their strapless shoes clip clop over the tiled floor surface. Both wear well-worn blue jeans and black leather jackets. Their long blond hair and black hair hang limply around their shoulders. They disappear.

Later...

A ubiquitous fleece smiles at me. Why? We are total strangers. I smile back. It feels good. I sit there feeling almost elated. A sense of goodness. I find myself smiling at all around me. I get strange looks. Eye contact avoidance procedures.

13 January

Have you ever noticed when people have a choice of TV screens to look at that they choose the one away from them rather than the screen nearest to them! Five people are all bending their necks to catch sight of the screen just above my head. It's an advert about buying clothes and books at reduced prices. You know the one, buy a hundred books and get one for £1.99. Not particularly gripping TV, but I suppose if you are refuelling yourself for another 100 miles of motorway before the next pit stop it does allow the head to wander a bit.

A father in his mid-thirties, casually dressed in high street gear, is sitting with his two blond-haired under-fives, a boy and a girl. The latter swings her legs under the table as Dad cuts up her food for her. The little boy eats contentedly watching the every move of his father. Dad is conscious people are looking at him. You can feel the absent father with kids on a Sunday afternoon thoughts.

*

Shopping: No matter how organised you may be, some things are always forgotten when you are away from home. BUT DON'T PANIC – Welcome Break sites, have everything you need from newspapers, drinks, books, gifts, sandwiches, music and much more – including exceptional offers on brand names.

*

15 January

Spend more than the statutory six minutes in cubicle 7. Mario impatiently pushes the mop under my door. I hear him puffing and cursing under his breath. I give him a sheepish look as I exit. He barges past me with his 'hi-tech' cleaning equipment on wheels. I hear the flush go proceeded by crashing and banging as the toilet seat and lid are thrown back. The mop swings into action on the floor. The hissing of the cleaning spray follows. Another 'customer' is temporarily barred entry as Mario blocks the narrow corridor to the higher numbered cubicles. I know he will be watching me from now on.

Later…

A ten-minute coffee with two senior Welcome Break staff.

'Over ten million people pass through South Mimms each year. Customers are counted by a 'fast counter' when they enter the slip road. Have you seen the camera? 60% of people visiting South Mimms come here for the first time. I think some expect it to be like a 'greasy spoon' café. Our market research has shown that over 50% of customers don't purchase anything. It's not helped because we cannot advertise its services and promotions. Service stations cannot become one-stop shops. We could not open a branch of Boots or Tesco here.'

'It's funny but we get more than fifty items a week left here by people using the motorway, including wallets, handbags, and car keys. Can you imagine it? Car keys? Where do they go? One of our table cleaners found £10,000 in used bank notes wrapped in a *Sun* newspaper recently. Never was claimed.'

'What about the "phantom driller"? Small holes, known as "glory holes", have been appearing again in the gents' cubicles in the intervening walls. Presumably so that men can spy on other men as they sit down or whatever they are doing in there. Costs us a small fortune in replacement panels. If you see him in your travels give one of us the nod.'

'Do you remember the time a human hand was found at the side of the road by the Truckstop? Turned out to be a hoax. A latex-covered plastic hand. The police were called out and part of the service station area had to be closed.'

'We have all kinds of people calling in here including Princess Diana with her two sons to get a Burger King one Sunday afternoon. That caused quite a stir!'

17 January

Managed to drive over a cardboard box full of empty Foster lager cans. A strong whiff of stale beer and men's piss as I exit my car.

Two men are playing cards. Stud poker I think. Three watching sixty-plus-year-olds with them have shaky hands as they manoeuvre their cups of tea to their mouths like serious hangover merchants.

Five women all the wrong side of thirty, all friends, are

meeting up en route to a health farm weekend break. I listen in to talk of leaving the 'old man' at home to look after the kids. One of the group makes an illicit call on her mobile phone. The others watch and giggle.

18 January

Two senior citizens are playing the gaming machines in the dimly lit Game Zone area. He wears a flat cap two sizes too big for him. A cigarette hangs limply between his lips. She sits across from him. Her red housecoat is too long for her; it reaches the floor as she sits on a stool gazing into the flashing lights of luck. Her cigarette is blowing smoke like a lorry going uphill. She wipes her glasses and waves at the smoke.

Other people cluster around the machines. A teenager, no older than thirteen, and a sense of desperation about him, eyes everyone as they walk past. I get the treatment as well.

19 January

A coachload of mouthy Londoners arrive; women on a work's outing. The group descend on The Granary counter. Kim from Hong Kong, a student, gets the order mixed up to a chorus of laughter from the 'girls'.

'What do you want, sis?'

'Said to Mum when I opened my eyes this morning, where's me cheque card?'

'Tell you what, go round everyone and get a fiver for the driver.'

'Why we queuing here?'

'Five big cappuccinos, please,' she giggles; the weekend assistant blushes.

'You see the price of those?'

'What are they?'

''Scuse me, what are they called?'

'Baguettes.'

'Never heard of them.'

The noise levels have increased. Words hover above the tables as everyone talks at once. Lots of 'oohs' and 'aahs'. One of the group is dressed in a sports top and trousers with white Reebok trainers – her shaven head is hidden under a baseball cap. She sits on the margins of a group eating a full breakfast. A couple of mobile phones have appeared; women's heads duck under tables in private conversation.

'It's the boyfriend,' one jokes.

'Didn't know she still had it in her.'

'Dirty cow.'

The group organiser is busy between tables, her black trouser suit and short reddy-black hair gives her that authoritative businesswoman look. There is silence as she speaks to each table, followed by howls of laughter and much head movement as she moves on to the next table. It's time to go. They move off in table groups. I hear the birds outside singing again or is it Welcome Break TV?

20 January

Two people in their twenties, he is white, she is black, eat two Ultimate Breakfasts. No conversation as their knives and forks chatter. She is dressed all in black including a black hat. He is Mr Ben Sherman with obligatory baseball cap and dressed all in blue.

A little four-year-old boy is excitedly running around the tables. He falls and bangs his head and cries. No one moves to pick him up.

A black woman sitting with her husband and two small children bursts into tears, her hands covering her face. A small crowd gather around her; she is distraught. One of her daughters with a Stars and Stripes headscarf looks away in embarrassment.

Two heavily kitted out police officers leave their breakfasts and walk quickly out, radios talking away. A police motorcyclist in black leather takes their place. He carries a radio the size of one of those field phones you see in 1950s black and white war films. You know the one where John Wayne is summoning up his troops for a defiant push across the German enemy lines.

A smart brown-suited man appears, glasses in one hand, sits by himself, and looks around.

Notice some of the people just passing through gawking at the 'sitters'. Do some of them ever purchase a tea, coffee or food?

Maybe they are convinced the seating is reserved for 'members only'.

The brown-suited man looks North African, maybe Libyan. He stands up, looks around again and moves off. I feel strangely uncomfortable. Fifteen minutes later he has returned with a black briefcase and a tray with coffee. He looks at me as he checks messages on his mobile phone. He meticulously folds a napkin and places it carefully under his cup, and catches me watching him.

23 January

Lunchtime hustle and bustle. Coach travellers, business types, 'passer-through' people, 'wanderers', and 'lost souls'. The noise level is reaching Concorde flying roof top level. Welcome Break TV is drowned out to a muffled whine. Seems to be one of those days when people decide to wear a specific colour. For some reason red is today's colour: red jackets, red baseball caps, red fleeces sporting a 'Big Crew' on their backs and a sprinkling of men in red socks, tucked nicely into polished shiny black shoes. An assortment of red sweaters and tops and a couple of red scarves remind me of a poppy field I once walked through in Flanders.

A couple of brothers eat burgers, regular fries and milkshakes. Amazing how versatile hands are, holding a burger one moment, fingering fries the next and grabbing a suck from a carton the next. Ever seen the chimpanzees' party at London Zoo? Do they still have them? Well, you know what I mean – eye contact moves from food to what's happening around them, rapid eye movement. The older one jogs to the toilets after emptying his last few remaining chips from the bag. The younger one remains eating his burger and licking his fingers. Is it possible to eat such a meal with a knife and fork? I feel a monkey noise emanating from inside me, a need to share the remaining scraps of food with the two brothers. I sip my coffee instead, smiling at a middle-aged woman who has been watching my hand and arm movements, fingers under my armpits. I smile back.

'Travelling far?' I enquire.

'Salford, my husband's getting the teas.'

I leave my table and exit on all fours with the occasional upright position.

Out into the car park. I have to negotiate a path through three guys, casually dressed, just standing around admiring a brand new top of the range green Land Rover.

'Excuse me,' I whisper, looking up into the sky at three seven footers.

'Sorry pal, didn't see you.'

I bet you didn't, you big bastard.

They walk off.

'Where's the motor?'

'Over there.'

I watch them give a nearby white Range Rover a quick inspection. One's holding a copy of *Parker's Car Price Guide*. The shorter one's earrings glisten in the sun. Much crotch-scratching as they stand around talking. A handshake. Both cars drive off in a convoy. Deal successfully completed.

24 January

Just gone 9 p.m. Men sit around like raindrops on a car windscreen. A few suited smokers get a last draw before coming in out of the car park. Secret whispers down a mobile phone.

A woman appears, artistic looking, wearing a black headband and wrapped in a long black expensive coat. She stands talking into her Nokia mobile phone, her violin case and bags sitting at her feet. She is getting angry, and folds her arms.

A 'Mr Burton' dressed middle-aged man, with a long ponytail (is it a hair piece?), winks at her as he walks around with his tray full of Burger King delights. His friends queue for late breakfasts.

Dunn and Co. Remember them? The high street chain for the discerning male. They're closed now, of course, and turned into a Nero's, Costa Coffee or a Starbucks. Well, the last man to buy clothes from them is in. Sporting a blue serge dark suit, with lapels as wide as the hard shoulder of the M25. A classic brown trilby hat sits on the table. His raincoat, complete with buckled belt, hangs over a seat. Dark, heavy NHS glasses sit on his nose. His bald head reminds me of a younger Donald Pleasance. He is reading *Roget's Thesaurus* and the *Daily Express* at the same time.

A crowd of 'clubbers' are meeting up. They sit in the far corner of the smoking section wearing high street designer casual wear, and all looking like sixth formers. One of the boys periodically lets out the call of the South American Spotted Nothura (*Nothura Maculosa*) bird. It echoes eerily around the hangar. He is ignored by the few of us in.

The late shift of table clearers is on. They look to be new. They're a vacant bunch who stand around staring in between cautiously clearing the tables.

A couple appear and a monkey call rings out.

'Shut up!' someone shouts out.

Silence.

I am tempted to answer the call with the older male Silverback gorilla reply. Decide against it. Catch sight of an older Asian male, can of Red Bull in his left hand, staring at him from the table. His attention moves. He eyes up the rear of a blond woman at the counter, whilst talking down to his mobile phone.

Mr Dunn and Co. returns to his table. His glasses retrieved from his top pocket he eyes us all up and down. I get a shudder down my spine.

I begin to feel encircled as six men all on mobile phones sit at various tables all around me. Are they speaking to each other?

26 January

Sanjeev greets me with a smiley 'Good morning'.

I tell you it makes a difference to your day if you get such a greeting as you make your way to work.

A group of boy soldiers eat their way through their breakfasts accompanied by Coca-Colas and Crunchie bars.

Long queues at The Granary this morning. Ultimate Breakfasts being eagerly snapped up by two guys in red and white jackets with 'Honda' emblazoned on the back, look as if they have just come off a North Sea ferry, dirty baseball caps, jeans which need a good wash and faces stained with oil and grease.

'Buy five at 99p and you get one free,' Paula informs her customer in the shop, her Portuguese accent hidden under recent ESOL tuition. A seventeen-year-old in with a party of friends shyly refuses the offer and blushes.

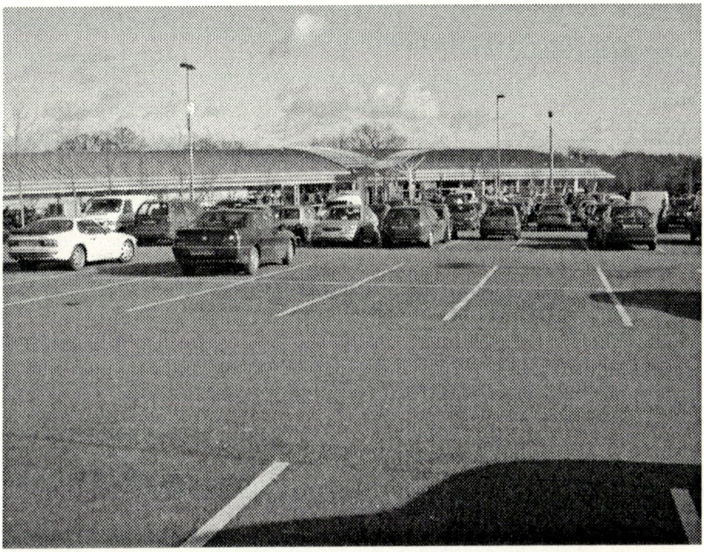

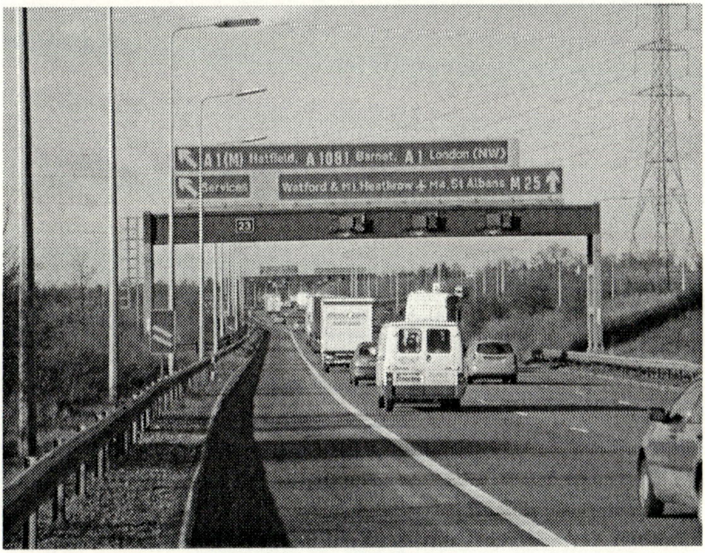
A1(M) Hatfield, A1081 Barnet, A1 London (NW)
Services
M25
23

29 January

An elderly couple, Mr and Mrs King, live within hearing distance of the motorway in a nearby village. She talks, he nods.

'We actually have only used the new service station once when it was first opened just to see what it was like. About ten years ago. We haven't been back since. I suppose it's too near home.

'Before it became a service station and before the motorway came, on Sunday evenings with my mum and dad, it would have been in the 1940s, so I was seven, we used to take an evening walk down Bridge Brook Lane, across the bridge, across the pig field, over the A1 to the Middlesex Arms for a drink. On the way we would pick blackberries. My brother and I would sit outside and have a lemonade. My mum and dad would sit and have a stout. The family weren't drinkers at all but it was just like an evening walk to pick some blackberries.

'The place that I used most of all was the San Marino swimming pool. That was an open air swimming pool. It was freezing cold. I think it was only open in the summer. It was in the middle of nowhere but in those days it was a place to walk and I suppose where we would walk from home, was probably a mile and a half through lanes and woods and fields to get there to swim. You had to dodge the A1 to get to it.

'I think it probably cost us 1s 6d to get in which was quite a lot of money in those days. They had cubicles which didn't have doors on, women one side and men the other. All you could buy was a cup of tea and a bun. There was a little café bit but it was only a hole in the wall type thing and it was up the bank, it stood up high, up the bank off the A1.

'It was the Barnet Bypass. Now the A1M. You had to cross that. No traffic lights or zebra crossing. Nothing. There wasn't a crossing there at all.

'That swimming pool was in a funny position because there were no bus stops or anything. I don't remember any bus stops or any access to it. I don't even know that buses ran along there. It was by the old Beacon Café, a lorry driver's pull-in. It was quite a rough old place but my dad used to take me and my brother across there. After we had had the lemonade, we walked back across the fields and he bought us ice cream because it was the best ice cream, it really was superb ice cream. Seems funny now it's all gone.'

30 January

Serious meetings all around me. Glossy brochures spread out on tables, laptops clicking away, pens exploring mouths, nodding heads, rubbing of hands, and ear touching.

A small child with her mother and grandmother has a balloon which bobs around much to the annoyance of an adjoining table of serious meeting people.

'Do you have five minutes?'

I look up. A woman, blond hair going on grey, with glasses covering a happy face smiles down at me. The butterfly brooch on her jacket lapel is almost hidden by a large clipboard.

'Sure.'

She sits down at my table.

'I am doing market research.'

I am taken through a list of questions.

'Average.'

'Very good.'

'Staff are friendly.'

'Six times a week.'

'Usually clean.'

Felicity informs me that there are a number of regulars she sees in here.

'People come in from the surrounding towns – Borehamwood, South Mimms, and Radlett. They come to eat: people doing shift work, train drivers or single people who can't be bothered to cook for themselves. A lot of home workers who live locally come in to sit here, read their newspapers and go home again. The idea that more and more people will work at home doesn't work. We need other people. See that couple over there?' she points a Stars and Stripes painted fingernail. 'They're regulars who meet up for a chat. If they opened a Post Office here people wouldn't go anywhere else but here. Well, must go now, I am dying for a cigarette. It's so nice to meet someone normal here, so many oddballs.'

I watch her disappear in the direction of the smokers' section. She never asked me what I was doing here!

31 January

A male Chinese peasant, with his wife dutifully six paces behind him, follows me in. Both look around as they move about the building fresh from the paddy field complex across from the car park.

A Finnish family, all with bleached blond hair, sit eating their Burger King meals out of the wrappers. Their heads bob up and down rhythmically as they gorge their meals. Dad periodically takes a break from eating and pans his camcorder around the building. He seems to have a particular fascination with the roof. He points it in my direction. I wave, he waves back, and then continues searching for the next subject for his home movie.

2 February

Listen to John Peel on Radio 4's 'Home Truths' driving in. A story about a father who didn't like children. So what's new?

A bonehead with an England logo on his sweatshirt instructs his girlfriend and a small boy (his son?) to wait outside the shop as he goes inside.

A Gerry Adams lookalike sits across from me dressed in his BBC interview clothes. He's engaged in conversation with two women both dressed in shades of purple and red.

The 'student' coach has arrived and emptied its contents – noisy, but articulate young twenty-year-olds in sweat tops, jeans and trainers. Momentary confusion as two 'senior' travellers have to readjust their stride and take their trays away from the laughter and crisps packets to a more 'grown up' area.

A cluster of people stand around the Game Zone area, talking, smoking, and waiting for friends and family to emerge from the toilets, before boarding a Wallace Arnold coach. Little boys are in their favourite football shirts. An elderly man in an old herringbone suit gently guides his middle-aged son to the toilets. The 'boy' has sticky-out ears, wears NHS glasses, and is dressed in a rambler's anorak. I am not sure who looks the oldest.

3 February

Push my way through a busy car park. Two people in a dirty Ford Mondeo grope with each other in the front seats. Both wear fluorescent jackets. An empty baby seat sits in the back.

A group of classic Land Rover enthusiasts are meeting up. Their accessories, large pedigree dogs of varying shades and breeds stand around barking. Children run around happily whilst the owners and their partners admire each other's waxed metallic bodies.

Coffee's good today. Two women in grey and off-red sweat-tops hold hands across a table. Their conversation is an animated whisper.

Catch a large man, physically indistinguishable from Jabba the Hut, playing the machines in the Game Zone, sunglasses dangle around his white tee shirt neck. His wife, younger, and much thinner, stands watching him, a collection of Sunday papers under her arm. People coming in slow down as they walk past to check if he is winning.

5 February

Mario, the toilet cleaner, is having a problem. A cubicle door is locked and he is on his knees looking under the door. He seems to be there for ages. Worried, he dashes out and returns with one of the African security guards.

Julius, resplendent in his high street security firm uniform, attempts to gain entry. His Pat Jennings sized hands fumble with a tiny key. I watch along with a couple of other people.

'It's locked from inside.'

They communicate with each other in pidgin English after quickly deciding that Julius's French and Mario's Portuguese are not compatible.

Julius is around to the adjacent cubicle and on to the toilet seat in a flash.

'No one here.'

Mario scratches his head.

He disappears and emerges from inside the locked cubicle.

The cubicle is empty. He smiles at the vacant space.

'Odd,' I whisper as if watching the opening of a robbed Egyptian tomb.

There are no climbing ropes, no visible evidence of an escape.

Mario pushes past him, mop and bucket in his hands. He has a thirty-minute cleaning schedule and is anxious to get on with meeting his targets.

Julius stands there arms folded looking into the cubicle. He is smiling as Mario explores the emptiness of the cubicle. 'It's always happening here.'

6 February

Drive in and catch a white guy in his twenties peeing into a bush at the side of the road. I dim my headlights to give him some privacy.

A camera flash goes off behind me. A photograph has been taken by a middle-aged Swedish-looking grandmother complete with pince-nez glasses. Her subjects are two younger women and an older man kitted out in Hebridean pullovers and matching twilled trousers.

Decide a couple of late sixty-year-old men in black baseball caps are harmless despite their glances at every woman who passes within eye shot.

My café latté has an unusual taste. Ben, who served me, I think pressed the wrong buttons so I seem to have a chocolate-flavoured tea!

7 February

Alan is an architect and lives seven miles from the service station. He has spent much of his life criss-crossing the South Mimms junction.

'I very rarely go into the service station. If I do it's usually only to have a pee if the traffic is bad, or something has happened, or if I know that I am not going to manage to get the office, I will sometimes go in there and have a quick cup of coffee.

'I would say now it's one of the better ones because it is so modern. It reminds me slightly of the old science museum because the aerial elements they have got stretched throughout the place looks somewhat like an early plane.

'The thing it suffers from is over-use, particularly when approached from the west. The government policy of introducing these shorter stopovers every ten miles, which I think is still the main government policy around all the motorway system, hasn't happened. Nothing has been implemented as far as I am aware and the positioning of the service station and its popularity lead to its own problems because it has terrible bottlenecks getting into it which stops all the traffic in all directions.

'The first recollection I have of the area is when it had traffic lights on it and it was the junction of the A1 and the A6 as it then was, the old St Albans Road. We used to go past it on a double-decker bus. The 84 bus ran from Arnos Grove to St Albans.

'When I used to cycle past there in the mid-1950s, I was about thirteen, in the evening from Friern Barnet to St Albans you could do that in about forty-five minutes. By the end of the 1950s and the early 60s when we were into motorbikes we used to have impromptu barbecues down at the bottom of Wash Lane close to the bridge (behind the hotel). Two or three times we were chased away by the police which was pretty pointless, as we weren't harming anybody.

'The Middlesex Arms pub at the time stood well back from the road and had a small green in front of it and I guess was only demolished in the early 1960s. Fundamentally what happened to the junction, as I recall it, was that it was a set of traffic lights that clearly had been there since the late 1920s or the early 1930s when the bypass was built. That was then replaced by a roundabout which had been built to the west of the traffic lights and that in turn was then replaced by half of the existing complex for a few years. Later the whole of the existing complex was built when the flyover for the M25 was finished.

'The M25 stopped there for a long while and, of course, that stretch of the M25 was the last stretch of the M25 to be built. So when the M25 was opened Maggie Thatcher went there, to the South Mimms area, and the ribbon was cut – I can't remember exactly where the ribbon was cut – but that was where all the press photographs were taken of her cutting the ribbon to finish the road.

'Even in the time I used to ride my bicycle up there in the early 1950s through to now you have got all these layers upon layers of old roads – the old London to Holyhead Coach Road which was superseded by the Barnet to St Albans Road. Much later came the Barnet Bypass which knocked out all of the lane system, and subsequently the M25 which only picked up very slightly on the stuff to the east. Of course it follows the old road to the west up and over Ridge Hill.

'I have worked in Barnet for thirty years and that equates to over 15,000 crossings of the South Mimms junction. If you take into account the previous twenty-five years that I have lived in the area I guess I have probably crossed it over 17,500 times which is slightly sad. I did mention that to my wife this morning and she said it's possibly time I did something else.'

9 February

Pull in to car park. Faces stare at me from a caravanette on wheels. A couple of 'grey heads' pop up and down to get a good look at me, their tea cups glued to their fingers. I smile and wave. Net curtains are quickly drawn across the window.

A whiff of cheese and onion crisps reaches my table or is it the lack of personal hygiene? I suspect the woman behind me who languishes in her seat facing a sandwich munching man wearing a flat cap.

Mum carries Dad's tea and coffee on a tray. He walks behind her looking lost. Sitting down, I am reminded of a mother and a son, except in this case they are both in their forties and husband and wife!

Go and sit by the exit at the La Brioche Dorée café. It's closed for refurbishment. Day trippers coming in look at me and go and queue at the empty counter.

'Is it open, mate?'
'Afraid not.'
'Where do we get a drink?'
'Over there.'
I point in the direction of where everyone else is queuing.
People stare at me.
Four men in black suits, white shirts and black ties – Last

Orders funeral gear – approach me. 'Can we get a coffee here?'
'Where did you get café latté from?'
I leave.

10 February

A granny in a red top and white hair clasps a baguette with both hands and vainly attempts to manoeuvre it into her mouth. Her young grandson wearing his new trainers and a David Beckham number 7 football shirt watches her with disgust.

Three wheelchair-bound adults circle a table before settling in with their café lattés, their path having been cleared by an aggressive pathfinder, a woman with short unfussy mousy hair.

Parents are bringing in their young sons for Burger King 'treats'. It's post-football match burger and fries for Jason, Tom and Michael. Expensive club mini-kits and replicas play two touch along the table paths. Budding David Beckhams, Joe Coles and Michael Owens run around without a football.

'Close him down.'
'Fucking hell, ref!'
'Offside? You must be joking!'
'Through ball, quick!'
'Get down those flanks!'

*

DAYS INN SOUTH MIMMS at SOUTH MIMMS SERVICES: Hotel facilities include meeting room, in-house movies, Sky TV, air-conditioned reception, optional room service breakfast and direct dial telephone/modem in every room.

*

14 February

Sunlight streams through the glass roof creating elongated shadows. Three nuns march in; people move out of their way as they head in the direction of the Games Zone area to play the machines?

17 February

A cathedral quietness, no Welcome Break TV blasting out. The congregation – sorry customers – move slowly from The Granary with their trays piled high with traditional Sunday morning breakfasts. A hushed reverence as the knives and forks are replenished. People talk in whispers. The occasional 'more toast, please' breaks the tranquillity.

A whiff of brown sauce floats over me. It's a Northerner eating her two-plate breakfast with extras.

German students stand around in small groups by the coach exit door. Some move off to the Game Zone, others venture into the shop. Both move in Panzer-like formations. They are polite and efficient as they plunder the shop. Refrigerator magnets of St Paul's Cathedral and the Tower of London seem to be popular.

Three Americans, complete with big mouths, take over the 'balcony' section. Regular punters and people travelling through run for cover as the decibel count goes up through the roof.

Sunday newspapers and colour supplements read over teas and coffees. Two Sikhs stand guard by the exit on their way to the Gurdwara, one in a black turban and the other sporting a Manchester City blue turban.

Later...

Watch an ever-increasing flow of people emerge from the sunlight into the 'cathedral'. It's disrupted only by the existing strollers and the usual people 'lost' and using mobiles to reconnect with their loved ones somewhere in the building.

A three-year-old girl dressed in a pink hooded top falls to the floor screaming. Her mother, smoking a cigarette, watches. No Burger King delight for her as people step over her and walk around this tormented soul. Her red Wellies stick to the floor as she is slowly dragged screaming towards the exit.

A customer announcement in nasal tones asks that Miss Pearce go to the customer manager's desk.

Sometimes the human form is so extraordinary that I wonder if we all live on the same planet or if some individuals, even whole families, have decided to take a break in their journey between solar systems and called in for a large latté and a doughnut! Clothes from unknown non-earthly retailers; hair styles only found in the Red Dwarf series; speech which muddles East End rhyming slang and railway station announcements; behaviour, such as 'side-walking', the result of a gravity-free atmosphere?

21 February

Manage to walk past the credit card sellers twice without being asked the now obligatory, 'And how many credit cards do you have, sir?' People, mostly men, are stopped in front of me, behind me, alongside me. I seem not to have that £7k credit limit look about me. Feel slightly left out; perhaps I should go over to the two young women and complain. Decide to leave it. Never fancied another credit card anyway (says the man with several already). Sit down with my large café latté, now my favourite drink.

'Excuse me, you have dropped your credit card.'

A Benny Green black cabbie is helpfully pointing to the floor – sure enough, one of my credit cards is lying face up on the floor alongside me – 'tis a sign. Well, if you can't get a new one why not throw out the old ones?

'Thanks,' I murmur, half smiling.

'Easily done.'

He gives me a worldly London cabbie smile which says, I've seen it all before, mate.

Jim, my mate, rings me on my new mobile phone. The businessmen around me nod their approval. I slip into business mode. Business is conducted. I have joined the club.

'Sure, we need a meeting.'

'Okay, let's go for it. Ciao.'

Business concluded I sit back with my 'new' business colleagues and try to look serious as I shuffle my papers and leave my mobile on the table alongside my latté. It feels good.

Julia is alongside me as I get my papers together to leave, her white-gloved hands are collecting my cup and saucer.

'Thank you.'

It's an involuntary gesture on my part which I always seem to do when confronted by people who clean up our mess who are low paid, work shifts, and in non-unionised jobs. She looks at me surprised.

Outside a black juvenile crow perches on the car. Trouble? No trouble! 'We'll get you started FREE' sticker on one of the lamp posts. I swear it's looking down at me and laughing.

3 March

The fifty-year-old male with black Elvis Costello glasses on a shaven head walks around not buying anything. He reminds me of an inquisitive dormouse.

A couple decide to leave after facing each other for the past thirty minutes in total silence. His diamond patterned socks help his brown shoes tap gently at her blue jeans. She hides her face under dark glasses. They have that illicit weekend feel to them. He has office worker's hands. Dressed in a golfing jumper and trousers which are 'over-creased', his hair is from a 'would you like something for the weekend?' barber, nicely brilliantined. They walk off side by side, not really a couple. I watch them walk to separate cars in separate parts of the car park.

8 March

It's approaching midnight. Spaces in the car park. People wandering under the lights. Small clusters of people gathered inside. The shops' grills are down, shelves are being restocked, chairs stacked on tables. The Welcome Break TV is talking to nobody in particular. Very few staff on duty, queues everywhere.

'Two slices, please,' I answer. She disappears behind the door marked 'Staff Only'. An Asian couple alongside me are also waiting for toast and a fried egg.

'Did you see that speed camera where the roadworks were?'

A coach driver with no passengers is talking to no one in particular. I nod. A number 2 Gary Neville white Manchester United football shirt stares at me from the back of a Malaysian lad with his three friends further up the queue.

A small clutch of smokers sit huddled together for mutual support away in the corner.

'You are watching Welcome Break TV.'

A number of middle-aged men sitting alone stare at the TV screen. All wear fleeces, have their arms folded and wear glasses. I check out myself, glasses – yes; arms folded – sort of; fleece – yes. I quickly remove my fleece, roll my sleeves up, and read my newspaper without my glasses on!

10: 'It's all happening here' – England v. Nigeria

> I drive for a couple of hours and then stop at a service station for a cup of tea and a doughnut.
>
> *Nick Hornby*

Grey heads bound for Scarborough – a passing customer – Germans in the gents' – bit of a kerfuffle – Bank Holiday disorder – Coffee Primo appears – a drive around the manor – tasty geezers – Mr Ryan – Café latté with Jim – cigarette run – World Cup fever.

11 March

The coaches bound for Scarborough are in. Lots of grey heads walking around with trays full of teas and cakes. 'Coachies' walk slowly and deliberately. They observe the speed limit of one mile per hour. Impatient queues form behind them like an urban traffic jam. The table cleaners stand around helpless as the grey heads sit rooted to their seats awaiting the collective call to embark. Teapots are long drained and only crumbs remain on plates, but still they talk and watch. Welcome Break TV strains to hear itself over the din.

Some of the coaches wear white trainers with slits cut to enable their corn-riddled toes to poke out for air.

Funny how people of a certain age hold their tea or coffee cup in their hand – slightly raised – and how they run their tongues around their mouths as if cleaning their teeth before the next swig.

One coachie is walking around carrying all ten daily papers under his left arm.

A couple of Chinese looking backpackers, 5-star hotel variety, stumble through the mass of geriatric humanity, looking bemused, their puff jackets and money carrying size rucksacks contrast to the local high street fashions around them.

'How are you?'

Des at my shoulder. He's one of the two staff troubleshooters who walk around picking up on customer's negative vibes before they get out of hand. He reminds me of David Jason's Inspector Frost.

'It's coach traffic, always bad this time of day at the weekend. It's a bit like the pubs closing – everyone comes out together. You here long?'

14 March

A passing customer

'I first came here a couple of years ago in 1997 for reasons I needn't explain. I responded to a personal ad by a lady living in North London. Let's call her Simone. She sounded fun. We wrote letters to each other. These were the days before e-mails took off. We got on really well. I phoned her and she had a sexy voice – very breathy, earthy, and seductive. From our telephone conversations she tells me she is married but needs a male friend. I decide to give it a go.

'Our letters get more frantic. She writes in block capitals, with many exclamation marks. Then, the crunch. Should we meet? Yes, but when and where? I suggest a couple of pubs not too far from her home. She demurred. She felt there might be someone there who would wonder who she was with. I suggested another pub, quite a way from her home. No good, she says. Her cousin goes there, and would wonder. We kicked it around a bit and eventually, and with some hesitation, I suggested the M25 service station at South Mimms. Yes, she said, that seems okay. She agrees to meet outside Burger King on the agreed day.

'So the fateful day arrives and I am feeling nervous. I get to South Mimms, park and walk to Burger King. Look around, no sign of Simone. I wait for thirty-five minutes but she doesn't come. So I go back to my car and ring her mobile. She answers, yes, I saw you, but I could not actually meet you because you look so old. I tell her I am not old, and you might have got the wrong person, so why don't you come back again? Because, after all, we do get on well, and it would be a shame not to meet wouldn't it? Okay, she says, turning on the charm. You might think, what had I to lose. I was genuinely curious about her. Okay, she says, she will drive back to South Mimms and be with me in an hour. I waited patiently. Then she appears. She is dressed in what looks like Oxfam charity shop clothes. Her hair is blond, long and very untidy. She has a wild and flaky look. Her voice sounds quite different now that I have met her.

'We sit here under the light so we can see I don't have any wrinkles. Order coffee and some cakes and she takes a sip. Then she tells me about a man in Devon who replied to her ad and she has seen him three times in two weeks. He drives all the way up from wherever, meets her in London, and drives all the way back. Then arranges to see her again. This is wonderful, she says.

'Now I listen to all this, wondering why is she telling me. Who is this date with? I ask myself. She bangs her cup down, tells me the coffee is lousy, the Devon man is terrific, and he's not old like me. Then she tells me why ever did I make her come to some bloody service station? She gets to her feet, strides out of the restaurant without a goodbye or thank you, and is gone. I am left with the bill. I ask myself what the hell was all that about?

'I still come here for a coffee from time to time. You never know who you might meet.'

19 March

A 'gentleman' appears wearing a Norfolk jacket, baggy corduroy trousers, and a pair of brown brogues. His plain grey socks peek out from beneath the one-inch turn-ups.

An Irishman with a County Claire accent orders a bacon and tomato roll. His mate, a Londoner, communicates with him in a 'Mid Irish Sea' dialect. I stand and listen. Reminds me of Brad Pitt playing the part of a traveller in *Snatch*. They both have hands the size of street shovels.

A beautiful businesswoman, with a brown briefcase, fawn mohair trouser suit, and long mousy hair. Men's heads turn a collective 360 degrees like a Jim Carey film. Nick behind the coffee counter says it must be the brown briefcase. I nod, spilling my large café latté down the front of my trousers.

20 March

Three large bare-chested German men in the toilets washing at the sinks, making monkey-type noises and yelps, mixed with raucous laughter.

*

'M25, Lazy Lenny warns us of an accident between the Potters Bar turn off and South Mimms, Junction 23. Thanks for that Lenny.'

*

21 March

It's Thursday again. Arsenal lost last night to Juventus in the Champion's League, difficult to feel sorry being a Spurs supporter myself.

Mobile phone ringing, a sort of *Upstairs Downstairs* tune.

'It's your mobile phone, Harry.'

'It's in my trouser pocket.'

'It's your mobile ringing, Harry.'

'Oh.'

Couple of grey heads from Lancashire, both small, overweight and jolly. Mrs Harry smiles at me as Mr Harry struggles to locate his mobile phone deep inside one of his Burton XL trouser leg pockets. I smile back sharing her 'it's lucky he's got his head screwed on' face. They share a pot of tea. Mrs Harry sits with a cup between her hands, elbows on the table.

Harry tugs at his brown leather belt, loosening it to allow for an expensive Welcome Break jam doughnut to ease its way down.

26 March

Half-term. It's deserted. A few lone businessmen are scattered around like some biblical seeds.

Bit of a kerfuffle. A Ray Mears type character in khaki with matching baseball cap and dark glasses berates one of the table cleaners. He has a dirty saucer – big deal. She is upset and is commiserated by her colleagues behind the screens in the 'box'.

'Ray' is marching backwards and forwards at double pace to pick up some serviettes. Other customers move quickly out of his way. A man with a mission, his coffee is consumed military fashion – 'Present Arms' coffee cup raised. Attention. He is up and gone to do battle with another service station somewhere else.

27 March

Four people sit together by the windows. A middle-aged white man, with a receding hair line, is in a last year's Christmas present – a black leather jacket. He's accompanied by three small Malay women, all five feet tall, one of whom is a stunning twenty-year-old with long black hair. The other two seem to be her older 'chaperones'. They talk together. The white man reads his English newspaper disinterested.

29 March

It's Good Friday, sunny with a deep blue sky.

How the great British public love to get out at Bank Holidays and enjoy themselves. It's boisterous, chaotic and ritualistic. A bit of seasonal disorder. ('The Bank Holidays were regularly associated with heavy drinking, shrieking girls, and the occasional free fight', Geoffrey Pearson, *Hooligan: A History of Respectable Fears*.)

Big breakfasts for the long road ahead, just squeeze out the last drop of urine before setting off in case there's no toilet on the way. It's the UK for Christ sake, not the Australian outback!

The regulars are noticeable by their absence. People who have never entered a service station complex before are everywhere.

'Oh, it's big in here.'
'Look, a shop.'
'Did you see those toilets, very clean.'
'Look, a restaurant.'
'Where are we?'

Groups of excited small children running amok and bumping their heads followed by floods of tears. Their parents watch impassively whilst their Darrens and Tinas drive their lethal three-stone bodies at other human beings until a crash occurs and then they are up out of their seats comforting their offspring. Meanwhile the rest of us breathe a sigh of relief at having narrowly avoided being run over.

Guy sitting in an orange tee shirt with a white number 20 emblazoned on the back. His blond hair is in dreadlocks style. He sits alone with a full plate. Has that passing through, leave no shadow feel. Eats out of the side of his mouth, leaves crumbs over the nice blue speckled carpet. Lucky my mum is not around.

It's a funny old day. Catch sight of three Welcome Break managers in the Game Zone playing a machine. Well, it is Good Friday! Somehow they are out of sync with the other punters.

A group of soldiers in battle fatigues have stormed the Burger King and are now sitting on the high stools eating in unison.

Get caught behind a twenty-something couple as they walk arm in arm to the toilets. A procession forms behind them as they stroll. Just one last kiss and they depart to their respective toilets. We all watch, bladders bursting.

I have a fifty-year-old woman sitting facing me, a few tables away. She is tucking into a large dish of quiche and assorted salad. Must have cost a bomb. Sunglasses rest on top of her ginger hair, her black blouse is unbuttoned halfway down, her large cleavage is prominent. She sits in a tight red skirt, with black stockings up to her bum, a pair of high-heel black shoes dangle off her toes. She watches what is going on around her in between liberal portions of food. Her mobile phone rings every few minutes and gives her obvious pleasure. She has an 'old brass' look about her. I am careful to avoid eye contact. A man walking past in light brown desert boots and an all-in-one matching pilot's suit complete with coloured epaulettes catches her eye. He has Arab features – Colonel Gaddafi meets Marlene Dietrich! Another phone call and she is up and off. A creepy guy in a black leather jacket like mine – worrying – eyes her up and down. She ignores him and tip taps out. Creepy guy unpacks his wallet and proceeds to download a couple of tablets, a swig of coffee and he is looking around again. Somehow I feel quite sorry for him.

The newly arrived Welcome Break 'El Supremo', is around. Stiff backs straighten, double pace, staff busy themselves.

A woman, not unattractive, is sitting close by the 'I am a lonely person but I am waiting for someone – anyone' table. It's literally half out of the doorway so is not really in or out. She sits transfixed, eyes watching the comings and goings.

People standing in a queue by the closed La Brioche Dorée counter again. What is it about the great British public? You form a queue and others join it; is the 'closed' notice invisible?

Mr Creepy leaves, his car keys swinging in his right hand. I notice he has no sideburns, just a pudding basin monk cut I used to get when I was seven years old which made my ears stick out like Black Adder.

'El Supremo' and two of his henchmen are checking out the fresh orange machine. It's not producing the required beverage.

Mobile phones spring into action, the blue light flashes, buzzers sound, staff running. The Burger King and KFC outlets are experiencing 'takeaway rage'. People desperate for their plastic food jostle each other, a baby is crying, a man, bald headed, shouts, a full tub of Pepsi is emptied on the floor – it's hell! I dive for cover.

22 April

Since my absence during the past three weeks there have been some changes. A new Game Zone has appeared next to the entrance/exit where the retail shop, Grand Prix, used to be. A Coffee Primo has also emerged complete with new coffee mugs.

Mr Huge Stomach enters. Not seen him for some time. He waddles in. A tall man in grey trousers. His hair, grey, is swept back. He has that leaning backwards gait so typical of overweight people who cannot see their feet when they walk but hope they do not tread on anyone.

Another pair of old favourites appear. The blond woman and her young teenage son with the hearing disability. They both stare in different directions seemingly oblivious of each other. The mother carefully pours the remaining cappuccino from the bottom of her mug. She is walking poorly this morning. Looks like a hip problem.

Mr Huge Stomach sits nearby. He carefully spreads his toast with butter and slowly manipulates his tea cup with his puffy white fingers. The next slice of toast hovers six inches from his mouth. A decision. The toast or the slice of bacon that has appeared from a side plate? Both are consumed at once. I hear the enjoyable munching at fifteen paces. Like a sleeping walrus, he rests over his copy of the *Daily Mail*. The occasional hiss of air as he sits contentedly. Have you ever noticed what tiny feet large men have? Almost baby size.

23 April

Danny greets me with a 'Good morning, Mr Roger,' and serves me my café latté and pain au chocolat without me putting in my order.

The mother and son with the hearing disability are in. They sit in the same seat as yesterday. Mum is wearing the same clothes: a multi-coloured jacket, a deep pink roll neck and mauve trousers. Her blond hair tops this outfit.

24 April

Mum and son are leaving. She is wearing the same outfit for the third day running.

Staff go about their daily business. Carlos is vacuuming the floor of the new Game Zone area whilst Des is patrolling the main thoroughfare. Chris and gang busy themselves with supplying coffees and bacon and tomato rolls to punters.

A couple of men walk about; one in a black leather bomber jacket, desert boots, and a number one hair cut; his mate, tall, shaved head, with a green jacket over a white tee shirt.

Body jewellery glistens as they peruse the public sitting at their tables. I am given a quick 'one two' lookover. They sit at opposite ends of the seating area, watching.

I try to pass off a Canadian two-dollar coin for another cup of coffee. Chris – eagle-eyed – catches me out.

I have a permanent bleeping in my right ear and muffled voices. Two heavily armed policemen are taking their breakfast break – both male. Their radios are left on in case of emergencies. Or is it just for bravado?

Why do people park so close to each other in car parks so that you have to lose instant weight to squeeze into your car?

25 April

A noisy lunchtime. A few kids with their families tucking into an assortment of takeaway food. So no school today?

Des is patrolling. Lost customers are picked up and slotted into the appropriate retail outlet. Pauline helps out a woman who needs a cab.

A boy is noisily playing the Fast Track 'shove' game in the newly completed Game Zone with his mother. The older brother looks on as he leans against his metallic crutch supporting his left leg.

26 April

I am driven around the large complex next to the service station by Kevin in his blue Mercedes. His elderly father sits in the front passenger seat. I sit in the back seat alongside a small aluminium ladder and a walking frame. It feels a bit like a Royal tour as people acknowledge our presence with nods and waves. There are lorries, buses, private cars, and crashed vehicles in allotment-size fenced areas.

'We took over the site in 1948. The Beacon, this building here, was built about 1931 and it was the first hostel for long-distance lorry drivers. In other words it provided a café for the daytime meals and then it also provided sleeping accommodation. We could accommodate about eighty drivers.

'It had been trading since 1931 and we bought them out in 1948 and then slowly we developed it. We bought this site where Scania are and then we bought the land at the back. That was a big farm and then we bought the site over the side there where the Truckstop is.

'In 1971 we had the Fire Regulations Act and it made the conditions of running a hotel the same as the thirty bob a night in those days we charged, as what you would have for a hotel which is thirty pound a night next door. It made it too onerous to carry on. No money in it. So we closed down the overnight sleeping because by that time anyway, they were in sleeper cabs and so things were changing.

'Now it's just offices there and a vehicle park. We have one hundred and seventy-five parking spaces at the back and there is also one hundred and twenty or whatever it is over there on the middle bit – that's the bit for the service area. That's organised by Welcome Break now; it used to be BP.

'We have other contracts like this one here, as you can see here, that are coming and going all the time, trucking in from Swindon, Glasgow, all over the place. This is all contract parking.

'Because we were established here as a commercial concern we had the diesel station but now that has gone with Welcome Break and they do all the catering facilities. We just do the contract parking here and basically all the lorries and coaches, apart from the coaches which go to the service station. There's a contract with Wallace Arnold, and the coaches go in that one and all the commercial stuff comes in here.

'There are most facilities here for the drivers although there's a pub down the road in the village, in South Mimms. You can't sell alcohol on a service area.

'We have developed the site and spent virtually millions because of all this concrete we put down. When the Ministry decided that this should be sort of a service area for the motorist and the commercial drivers, both Hertfordshire and Hertsmere councils said not a square inch more land than what is already there with Bignalls and ourselves. So what we have been trying to do is get a quart into a pint pot and then they moan that night-time vehicles are parked in the road. I said that's not our fault, we can't hang them on hooks. The point is that the whole idea of providing these places was to stop the vehicles, these heavy big juggernauts parking in residential streets; this is what the whole purpose of these service areas is, for lorry parks.

'You have got the motorway that side, the M25 is there so you can't go anywhere there. You have got the A1 there – you can't really do anything else. We actually own the land down the side of this, through the village there. All that field, that bungalow there, all the land along there. Another twenty-five acres there but, of course, it's the South Mimms village envelope and the planners won't let you do anything on there. But where this is, with the A1 there and the M25 there it can't go anywhere else really and you have got a little brook running down the bottom there. The present facilities are absolutely saturated.

'The Middlesex Arms pub was just across the road there. Of course that has all been knocked down. We used to go in there for lunch. It was nice.

'We have never been in the service station next door. Not since it was built.'

31 April

One of the pleasures of arriving at a service station on a motorway is the queue to get into it. For some reason this morning it has taken me some forty-five minutes to come off the motorway and park my car. Why does it take so long to come off the motorway and travel half a mile around to the service station? I do not know. There are no visible signs of an accident, i.e. people standing on the verge beyond the hard shoulder, nor the auditory 'dahs dahs' of police cars or ambulances. It's one of the great unsolved mysteries of motorway exiting to a service station. Personally I think there has been a rush for coffees or Fat Boy breakfasts and the service station management are only letting vehicles in twos and threes.

9 May

A maroon G registration Ford escort car has been dumped at the side of the car park entry road. It has a 'Police Aware' sticker on the driver's side window.

Arsenal supporters everywhere in their red shirts from various seasons. Takeaway food is being eaten by the bucket load. Looks like a Mad Hatter's Tea Party! Today's uniform consists of short hair, trainers and jogging bottoms. A few children and women are with the men. They all sit near each other as if on the terraces – no uniform chants, just collective mouthing movements and burps.

11 May

The car is still there; it now has all its windows smashed in.

Couple of tasty geezers, tattooed 'I love Mum' arms, in greasy sheepskin coats. Sledgehammers in the car. Bald heads shining in the early morning sun. A couple of large foaming Rottweiler dogs eyeballing me from behind the Land Rover windows.

At the other end of humanity a coachload of Japanese tourists has descended on the Saturday morning tranquillity. Standing around in large groups, ignoring queues, they talk excitedly behind their large-rimmed glasses. Other customers are sidelined. One group, led by a man with a voice loud enough not to use a megaphone, make a rush for the coffee counter. Perry and Lisa are overrun. Cries of help go ignored as one supervisor panics and dives for cover behind the 'Staff Only' doors.

As I leave, two soldiers come in wearing green fatigues. One has a fresh love bite on his neck.

14 May

The car has lost its wheels. Someone has had a good weekend. Crop circles of burnt tyre rubber are everywhere in the car park.

A swarm of American schoolchildren with assorted teachers and parents are in the shop. A long snake of customers queues patiently at the cash desk. A man with a single-tone Welsh accent exhausts my hearing and concentration capacity as he drones on to two colleagues.

15 May

Two of the cars doors have disappeared overnight.

16 May

Mr Ryan is in his seventies and lives with his wife in nearby Barnet. He remembers Bignalls Corner as a child.

'My father bought this field from the farmer, Mr Gibson. I think he rented it. He had a bungalow constructed on it right at the junction with the then Barnet Bypass and what was then called the new road which went from Barnet to South Mimms. Where it crossed the Barnet Byass there was a concrete base with a pole sticking up and a red light flashing in four directions to warn people of the crossroads. Of course it used to get knocked down by lorries and cars. I don't know why but it always seemed to finish up; we were much lower than the road, in the driveway of the house. This thing used to roll down and we used to find it finished up outside our front door. There was no traffic. Nothing. Might see a car every ten minutes.

'Diagonally opposite us was this garage, hardly more than a tin shack type of thing, which had one petrol pump which was, of course, with a handle, and that was Bignalls garage and that's how it got its name. Of course after that they developed nurseries and things like that. The bungalow became the Five Bells Restaurant and it was extended after when my father sold it.

'Where the café was at the corner of Wash Lane, where Dears, the transport people, came, my father built a little sausage factory on the corner of the field. It was there until, I think, about twenty years ago. It manufactured sausages and cooked hams.

'I lived at the bungalow until 1931 then came to live and go to school in Barnet.

'Before the fire there I went to the service station. It was a fine sight, of course, with a number of people passing through it. Haven't been back since. Don't know why.'

18 May

The car has been set alight and is now burnt out. Bits of it lie across the grass verge and the road.

*

Custom and Excise officers and armed police swooped on South Mimms services last week to seize 180 kg of cocaine from a Spanish lorry with an estimated street value of £12 million.

*

20 May

Meet up with my mate Jim for an early evening café latté and a chat. I catch him paying for a pair of pants at the shop checkout.

'You can't be careful these days. Never know when a spare pair in the car might come in handy.' I nod approvingly.

At the Coffee Primo he decides on a pastry – the Panata egg custard tart. They come in a pair and cost the equivalent of an hour's wages for someone on the minimum wage. I get little change out of a tenner. I can picture Sherpas bringing the pastries up over the high passes from faraway places such as Barnet and Potters Bar. As we talk, Danny, a Brazilian member of staff recently promoted to serving in the shop, hence the dark suit and collar and tie, comes over to where we are sitting.

'Hello, Mr Roger, how are you?'

We exchange pleasantries.

Two coffees further on a professional colleague of Jim's walks past en route to the toilets. Thirty-seven minutes later he re-emerges with his trouser belt flapping, followed by a younger man. They leave in convoy.

*

At South Mimms the road turns southward, at inclination of 1 in 22, it then bends considerably to the eastward, and descends in a crooked manner, at 1 in 25 to Mimms Wash, along which it passes for a considerable way upon so low ground as to be several feet underwater whenever there is a flood in the brook. (House of Commons. Sixth Report from the Select Committee on the road from London to Holyhead. Appendix: St Albans to South Mimms Trust, 1819)

*

27 May

The car has gone.

Cubicle 10 has a plastic bronze coloured women's size 16 plastic coat hanger on the back of the door.

'Charlie, you in there?

'Billy?'

'Which one you in?'

'Over here.'

Footsteps.

Knocking on a cubicle door.

'How long you gonna be?'

Silence.

'Charlie, you in there?'

Silence.

'Billy?'

'Where are you?'
'I'm coming.'
'Okay.'
'Yeah.'

A group of very young soldiers, fortunately without guns, are regrouping by the toilet entrance. Names are being called out.

Car park. A battered grey French Renault people carrier with no tax disc. Seated inside is a white guy in sunglasses lost on a large cropped blond head. He's wearing compulsory dodgy sports gear, an Adidas dark blue tracksuit with white stripe. He chain smokes. The vehicle is full up with cartons of cigarettes and parked next to a 760 Turbo Volvo. Inside, a black guy, in dark glasses, tries to look cool but doesn't quite manage it. His white blond female companion looks seedy. Their car has no back window. The three of them sit talking and smoking like teenagers on a street corner.

28 May

Scrawled on one of the toilet cubicle walls in black felt tip pen: 'Give us back our holes. We will shit behind all these panels.'

The silver metal panels which have been bolted onto the lower part of the dividing cubicle walls were introduced in a vain attempt to stop people from drilling through the walls and watching other men and boys sitting on the toilet!

12 June

Return from holiday. Watching England play Nigeria in the World Cup. There's a hushed silence about the place as all eyes are on the six television screens showing the match live.

Customers rush to their seats carrying trays full of teas, coffees, and plates piled high with cholesterol breakfasts.

I spill coffee down my shirt as I crane my neck forward and upwards at the TV screen suspended from the sky above.

Get into a football conversation with a sales rep.

'We're giving them too much space.'

'Bit of pressure needed.'

The Portuguese table cleaners are clustered around making

muted comments about the match. A Scotsman sits in front of me, silent and thoughtful. Mobile phones are going off everywhere.

'I'm at South Mimms, be there about 9.30 p.m. It's still nil nil.'

Some of the men watching the match don't look like football types, more 'soccer' sorts, who would be more comfortable watching the tennis at Wimbledon.

'Colin, it's Nigel. Where are you?'

'Stuck watching the England game. No goals, I think. It's exciting.'

A huge intake of breath, bit like a plumber's estimate of the cost of work needed to be carried out in your bathroom, as Nigeria miss a goal chance. Seaman drops the ball, a plate crashes to the floor behind me. Owen almost scores at the other end, a cup bounces off the floor, the TV screen temporarily freezes.

'Fucking hell, fucking typical!' People are up shouting.

Trevor Brooking's commentary is driving me mad. His nasal Chadwell Heath tones cut through the tension with all the finesse of blunt scissors.

A man in his forties, glasses the size of Cambridgeshire, sits right beneath one of the screens. He eats with his head tilted at forty-five degrees.

The match is a bore. I feel as if I am at home sitting in my front room on my favourite comfy chair as 'Mottie' feeds me more facts about the size of the pitch and how many bricks were needed to build the stadium.

People are getting edgy. No goals. Julie and Carol are trying to cope with the rush to their counter as people replenish their stock of fluid during a lull in the play. Des patrols the proceedings like a referee, neatly pressed black uniform, pens in his top pocket. He discretely intervenes at the first sign of trouble. An exchange of words. No yellow card this time.

Half-time arrives. Decide to make a dash to the gents' along with several other men. Hands go unwashed as we quickly return. Grab a coffee, but no meat pies to take back to the terraces. Seems some of the regulars have crept in during the half-time break. One of the African security guards, Julius, shares a joke with a shaven head television star. What's his name?

Another cup and saucer crash to the floor as the teams come out for the second half. Sol Campbell soon makes a forced error, we jump up, shout. There's a scream as a woman is on her feet. A couple of mobile phones spring to life.

'It's all happening here,' John Motson announces.

Two women, a mother and daughter, stand in front of the television screen to cries of 'sit down', and 'oi'. They dither; their Brasher walking boots leave footprints in the carpet pile as one of the table cleaners waves them over to an outlying table. Some of their friends have arrived. Overdressed for rambling across the fields of Hertfordshire, their rucksacks are carried through the football crowd. A group of them have sat down at the table directly below the television screen. They are oblivious to what's going on above their heads and the sixty-two pairs of eyes looking in their direction.

Two builders have joined my table – John and Steve. They're from South London, Battersea, and are not slow to get into the action.

'You fucking cunt, Beckham!'

'See that?' as Beckham puts a free kick over the Nigerian crossbar.

'Didn't I tell you he's a wanker?'

I nod politely.

Des clears our table. Nice to see a bit of personal service, although I catch sight of a red card just lying in wait in his top pocket. He eyeballs John and Steve. I expect the worst – my table down to ten men if John is given his marching orders. Steve returns, his blue baseball cap bobbing up and down in tandem with his carefully cultivated belly. Two Wembley-size Cokes and three bacon rolls are plonked onto the table. Sheringham misses a chance. 'God Save the Queen' is being sung by supporters in the stadium; the hairs at the back of my neck rise to the occasion. Tony, one of the sales reps alongside me, is mouthing the words. I am beginning to feel a bit tearful. We could be through to the next stage. Cigarette smoke drifts across the table as both John and Steve are puffing away like fifteen-year-olds behind the school bike shed. It's a non-smoking area and Des is approaching the table with a yellow card in his hand.

One minute of added time left to play. Some people are standing up; there's some shouting behind me; Argentina are drawing; they're out of the World Cup; people are cheering. They're out and we're through.

'The Argies are fucking stuffed, Rog!' Steve shouts in my direction.

It's a draw. Handshakes all round, a few hugs. John lets go a Coke belch.

It's all over.

11: Preacher Man

I've seen things people would not believe.

> Roy, in *Blade Runner: The Director's Cut.*

Offers on the toilet wall – Car Park no exit – 'Are you Roger Green?' – the Cambridge Don's wife – text from Jim – save the sinner – Mimms Wash surprise – haircut at Willie's – the army practise their shooting skills – caravan man – Mr and Mrs Atkins play the machines – car rally – being watched – three women the morning after – Mrs Allford's lorry rides – David Seaman? – last supper – Mark likes it here – get invited to a wedding.

14 June

'This is a customer announcement. Will the owner of a red Audio registration number xxxyyyz please go to the car park immediately and meet security.'

Three men from different tables proceed to the exit.

A gaggle of blue-robed nuns sweeps through the concourse. All wearing glasses, the sandled foursome walk two abreast to their destination – the toilet. A Ford Galaxy people bus waits patiently outside in the early morning sunny drizzle. Its blue bodywork matches the nuns' robes.

Another day, another wall. The cubicle wall I stare at has graffiti which offers me the alternative of either supporting the action against the 'puppy killers' of Huntingdon Life Sciences, the animal testing laboratory, or coming back at 1.30 p.m. today to meet up with a cock-sucking 26-year-old white male in cubicle 16.

15 June

Bit of a squeeze getting into the car park. A chauffeur-driven new black Jaguar is attempting to exit the car park at the entrance. It's one way. Simon, a Welcome Break employee, is waving and finger pointing at the driver. He's ignored! I end up facing him. His non-smiling face has a clipped moustache that is reminiscent of a Second World War RAF pilot. Bryl-creamed thinning black hair is fronted by rimless steel glasses.

'Tally ho, chaps.'

I notice he is gesticulating at me to move out of the way. The passenger's puffy soft white hand rests on the open left rear-side window. Fingers are tapping.

We eyeball each other.

A line of cars stretches out behind me. Should I pull over? His car horn lets out a heavily amplified blast. Several more follow from an impatient suit three cars behind me. A conversation is going on in the car and we sit stationary. He reverses up the kerb and lets us all pass before roaring off. I catch sight of the car in my mirror rushing to join the traffic queuing for the motorway. I pull in to park as a grubby red Nissan is exiting the car park against the flow of oncoming traffic accompanied by much horn blowing.

I go for a coffee.

A table cleaner carries a man's grey corduroy jacket away from one of the terrace tables. A mobile phone in the inside pocket is playing the theme from EastEnders. I look around for Pat Butcher.

*

GAME ZONE: While you relax from the motorway journey, the kids can entertain themselves with all the latest interactive and computer games, including Change Machines for your convenience.

*

17 June

It's hot. Men are asleep in their parked cars with their mouths open trying to catch flies.

'Are you Roger Green?'

I look up from my paper. 'Yes, who are you?'

'Colin. I was out in Romania with you in 1990.'

'Sorry, but I don't recognise you.'

'I was one of the drivers along with Graham. Do you

remember Kate? I married her. We have four children now. The youngest is four. She was a physiotherapist on the trip.'

It's 2.33 p.m. I am plunged back a decade to the grimy world of post-Ceausescu Bucharest where I worked in a children's orphanage with a small UK based charity. We were one of the first charities out there. It was tough going for all of us who went.

'So how did you recognise me?'

'I was sitting over there with my colleague and I kept saying to her I was sure that's Roger Green sitting over there who I was out in Romania with. So I came over.'

I join him and his colleague, who turns out to be a doctor, at an adjoining table. My table is swiftly cleared.

We chat about our time out there, what we are doing now. His colleague smiles in a polite but disinterested way as we reminisce. 'Frank the plumber', who woke up one morning in downtown Bucharest instead of High Wycombe, is remembered for his heavy drinking, his fondness for the occasional puff as he was soldering pipes, and his ponytail. Where is he now?

A group of pensioners on the next table earwig our conversation as they wait to board a coach.

'It changed my life. Probably did for all of us.'

Handshakes all round as he and his colleague depart to continue on their journey to Northampton.

I get another café latté and smile at an elderly man wearing odd shoes who asks me if I am Paul the coach driver.

19 June

Mrs England is nearly eighty-four years old and lives in South Mimms Village. Her late husband was a Cambridge Don. She thinks she is the oldest inhabitant in the village.

'The end of Greyhound Lane is completely changed because I always tell people that I remember that when I was a child we used to come down on the bus from Barnet at night and there were no lights, nothing at all. The bus used to stop at the Green Dragon Pub. When you think of it now, that service station and the enormous roundabout to what it was like, it was just fields.

'We used to have a meal at the Middlesex Arms pub down the bottom here; that just vanished with the motorway and then there was a transport café called The Rest. That was straight across the road from here. I remember the swimming pool. I didn't use it because I don't swim but I remember it yes, the Rainbow it was called. Was it the Rainbow or the San Marino? Oh yes, it became the San Marino.

'I had my twenty-first at the Five Bells Restaurant down there! That was during the War. They used to do wonderful ice cream, I remember. That was a long time ago! It was very pleasant. We walked down the lane. It was very nice down there.

'A friend and I often used to go and eat down there (service station) but you know there was the great fire. I don't think it's so nice since the fire. I liked it very much. I used to like it before because you went in and you went right through and you know Wash Lane. Well, it looked out on Wash Lane and it was so nice, but now it doesn't seem so nice to me.

'When people hear you're from South Mimms, well, now they say "Oh yes, that's where the service place is, the big roundabout." That's all they know about South Mimms is the service area because it was the first one on the motorway. The village has been forgotten.'

21 June

Get a text message from Jim: <At South Mimms. Buying more underpants. The first ones went down a treat.>

22 June

Two Italians come in: a bearded male and his blond girlfriend. They stand, look up and stare at the roof as if it's Michelangelo's Sistine Chapel.

Later…
Get asked to buy a raffle ticket by the Rev. Gerry Brown, a Pentecostal minister in a wheelchair. The prize is a 'Dinner and West End Show for two'. He has no legs.

'Lost them running to catch a train at Watford Station some thirty years ago.'

I nod, not sure what to say.

He reminds me of a friendly uncle. Dressed in a dark brown safari shirt his eyes smile through his rimless glasses. He slips me a recruitment leaflet for his church after I purchase a couple of £1 tickets. Mark, his minder, follows him around in between eating burgers and sucking a milkshake noisily.

Bored out of my head, I politely listen to Gerry as he goes through his routine to convert me. He gets the message and eventually wheels off to confront an unsuspecting Burger King family at another table.

*

Next time you find yourself having a cup of coffee and a stale Danish pastry in a motorway service station, look around you. What are those men with briefcases doing at the next table? Are they travelling salesmen comparing double glazing brochures? Or are they tycoons stitching together a secret deal?

Forget the Savoy Grill, big business these days is done in motorway cafés... The service station is the ideal solution, handy yet anonymous. Everyone else is too intent on bolting their snack and getting back on the road to wonder what fellow travellers are up to.'

Lucy Kellaway, Financial Times

*

23 June

I am sitting outside on the terrace enjoying my café latté. It's a summer morning. Facing me to the right is the entrance to the Days Inn Hotel where a grey-haired businessman is furtively making his exit. A younger woman follows discreetly behind him. They go to separate cars.

Ahead of me, almost within smoking distance, and hidden behind an uncared for hedgerow, lies Mimms Wash – a picturesque stretch of water meadow with the now disused Wash Lane weaving its path through it. Long ago this was on the old London to Holyhead coaching road. Mimmshall Brook, which runs parallel to the road, has been prone to seasonal flooding for centuries. So much so that, until a bridge was built in 1772, coach travellers had to disembark and cross the brook themselves on foot whilst the coach-driver and his horses found their own way over.

Not surprisingly, it became a convenient place for highwaymen to ply their trade, and in the late seventeenth century, highwaymen made off with £15,000 of collected taxes being brought down from the north. Some were not so lucky and a highwayman named William Ward was shot dead in the late eighteenth century and is now buried in South Mimms churchyard.

Climbing over the stile into the now overgrown Wash Lane, I bump into a couple making love on the hump-back bridge which is within viewing distance of the ground floor windows of the Days Inn patrons. The woman's legs are wrapped around the man's waist as she sits on the bridge parapet with him facing her. I cough loudly as I approach them. They seem oblivious to my presence as I quickly walk past with my head at a ninety degree angle turned away from them and concentrating on the rubbish which has been thrown in to Mimms Brook which runs under the bridge. Wanting just a quick peep to check if I wasn't imagining this scene, I look back and manage to stumble down the bank and end up doggy style in the brook with tepid water up to trouser tops. Still embraced in each other's arms I see them watching me as I pick myself up and squelch off up the lane in the direction of the motorway.

24 June

Haircut at Willie's Hair Bus. It's midday. Mick, a trucker, is getting a No. 1. Their mutual friend, Lee, has been sighted recently.

'He's working over at what's-it services. You know the one at the start of the M1.'

'Not sure.'

'Up to his usual tricks then?'

It's my turn.

'So how's the book going?'

'I went over there (service station) after you last came. What a place. Cost me a week's takings getting a coffee and a cake. Nice though.'

'What will it be then?'

We negotiate a No. 4.

He bobs and weaves around me, talking all the time. Periodically he disappears behind the curtain up the other end of the bus. The sound of fingers hitting a laptop can be heard.

A couple of truckers call in.

Willie doesn't rush. He has that skill of extracting information from you without you realising. A passing lorry drowns out my confession. We get into languages and he tells me he spoke Polish until the age of five.

'I speak English now although you wouldn't know it.'

The truckers laugh. He's playing to the audience.

27 June

Driving off the roundabout this morning and on to the service station slip road, I spot a man in a Tottenham Hotspur football shirt, blue shorts and matching football socks standing by the side of the road with his arms folded as if he is waiting for someone or something. On his face is a full-blown black Zapata moustache. Is it real? Has the team bus left him by mistake en route to a game? Or has Glen Hoddle dropped him?

In the car park a security guard in his uniform is sitting in the open boot of his once white Ford Fiesta car changing in to black regulation security shoes and white socks. Two traffic cones peek out of the boot from behind his back.

28 June

12.13 a.m. It's empty. Chairs are stacked neatly on the tables. A cleaner, constantly yawning, mops the floor of the Game Zone. His yellow trolley follows him. Elton John is singing 'What does it take to want me' to himself. Two Chinese people, male, female, sit at a corner table eating and talking excitedly. A shabby middle-aged white man is tucking in to a full English breakfast. Three armed police officers and myself sit and look at each other. A lone smoker views the proceedings from faraway in his designated section.

29 June

The army are in. A group of young soldiers, just into shaving, walk around dressed in their desert uniforms. They aimlessly drift into the Game Zone with bottles of Irn Bru in their hands. Shooting skills are practised. Others violently crash their cars on imaginary roads. One of the group, tall, with closely cropped hair that matches his temper, swears loudly as he jeers at his friends' feeble efforts. Julia, the only female soldier in the group, is treated as one of the lads. Her brown, almost shoulder-length hair hangs limply over her pretty suburban face. Mark challenges her to a game. She beats him. Berets are tossed in the air. She leads them out. They carry white plastic bags full of the *Sun* newspaper, bottles of soft drink and bars of chocolate.

Later…

The light green and white two-berth Swift caravan is still parked in a corner of the car park. Alongside it sits a nondescript red Vauxhall Astra saloon.

A slightly built white man, early thirties, in blue jeans and white socks only, moves between the caravan and the car. Both arms are covered in tattoos. How long has the caravan been parked up? All week? Does he have to pay a daily parking charge after the first two free hours? Who is he? What does he do?

The van's curtains are drawn, but the door is slightly ajar. I knock and enter. It's dark inside.

'Hello.'

'Yea, who the fuck are you?'

'Oh, I'm sort of writing a book about South Mimms Service Station. And was wondering if I could talk to you…'

He's up and at me. I am shoved out of the van backwards on to the bonnet of his car.

The van door is closed quietly.

1 July

Gents' toilet. Cubicle 16.
'Hello.'
'Yeh.'
'Really?'
'What sort of guy tells someone twenty years younger than them he loves them?'
'Plonker.'
'He'll get another heart attack.'
Laughter followed by a bout of sustained coughing.
'Sorry, brought the wrong fags back from Calais.'
'That makes me very, very happy.'
'Still up for next Thursday?'
'Oh, you are?'
'You're a gold star.'
'Have a lovely time.'
'Give her one for me.'
'All right.'
'Cheers, mate. Bye.'

2 July

It's Friday evening. Been a long week. I sit staring into space sipping my coffee. Excited young children, still in their school uniforms, run around. Lone men eat takeaway Burger Kings. Others tuck into meat and two veg. Danny at the Café Primo counter greets me. We shake hands. He's on late shift. We wish each other a good weekend. A late business deal is going down noisily behind me.
'Twenty thousand doors?'
'Jesus.'
'We can do that.'
'When by?'
'Next week.'
'Looks good.'
In the Game Zone, 'Over 18s only' say the signs, Mr and Mrs Atkins are playing the machines. Mr Atkins wears his lucky dirty black baseball cap. He sits in one corner. His wife sits near the

entrance crouched over the '25 carrot Gold' machine. She can play and watch passers-by en route to the toilet. She chain-smokes. Fifty-pence coins are repeatedly fed into its hungry mouth. The necessary three carrots required for a win do not materialise.

'We come here every night of the week. Well, it's somewhere to go. Nothing much else around here is open. This is open twenty-four hours a day, seven days a week. What you doing here?'

3 July

Four bright yellow Seat Leons are parked next to each. Another nine Leons from flash red to black magic metallic sit behind them. Their owners stand around talking and smoking. No children are present.

Not sure where I am. People sitting and waiting about, wearing summer beach clothes: Bermuda shorts with large pockets, Hawaiian style tops, and sleeveless 'I pump iron' vests – two-week sun-tans from facing the sun from morning till dusk somewhere on the Costa del Sol.

4 July

A 'face': he's a white male in his early fifties wearing a black three-piece suit which looks expensive but doesn't quite fit with his cheap, laced shoes from a local Matalan store. His dyed slicked back hair is in the style of a city merchant banker. I catch sight of him strolling around the tables looking disinterested but clocking everyone present at the same time. He vanishes to return with a half-filled glass of orange juice and sits in one of the corners enabling him to see who comes and goes. He folds his arms impatiently and looks over in my direction. I have the sensation that he has been doing this for some time, perhaps trying to make up his mind by close inspection if I fit the bill. The table clearers are oblivious to it all.

6 July

The wind is blowing outside like the great storm of 1987.

A cowboy follows me in from the car park complete with Stetson hat, Levi's jeans, and brown boots.

Inside, three scantily dressed young Irish traveller women are sprawled out on the seats in the Coffee Primo area. They have been given yellow fluorescent tops, the type you see hole diggers wearing at major road intersections in London as you watch from your car in the traffic jam. The tops are to cover them up. Two older, much tougher looking women are chaperoning them. Welcome Break managers patrol nearby speaking into their mobiles. A middle-aged man who looks the worse for wear joins the group. He sits quietly in the corner not saying a word as the women around him talk and shout at each other. Security are called. The women are asked to get up from the seats and move to the table and chairs section. A few expletives punctuate the air. The security man smiles back through his shiny white teeth. The group move slowly then run barefooted outside into the car park and scramble aboard a battered blue transit van marked 'Discount Paving. Call the Specialists'. The old boy with the group follows them out. He smells of whiskey and walks like a Donegal drunk on a Saturday night. The African security guard eases him out through the exit doors.

7 July

Danny comes over as I sip my coffee.

'Good morning, Mr Roger. I am off to Turkey to propose to my girlfriend. It's my birthday today so seeing you as well has brought me luck.'

Vic is standing on his wooden ladders listening to our conversation as he cleans the inside windows. Balancing on one leg he watches people as he goes through his graceful routine with his cloth and squeegee: no drips, no splashes of water. He's a professional. My vegetarian breakfast fry-up tastes good as I watch the free floor show. I am sure I hear a clap from one of the tables somewhere behind me as he completes his task.

Back outside seven BMW C1 motorbikes stand with each other minus their owners.

8 July

Mrs Allford lives in Hatfield just up the A1M from South Mimms. She drives around her ex-council house in a motorised wheelchair. Now in her eighties, she took a Social Science degree at Liverpool University and then went to live in London in the early 1940s. She remembers South Mimms before the service station.

'Between 1943 and 1950 I used what was then the Beacon Café at South Mimms as a pick-up point for a lift with lorries to Liverpool and back. It was a typical transport café, not all that large with the usual scrubbed wooden tables, bacon and eggs, and sausages and beans, and fried bread.

'They had very large lorries and very nice lorry drivers. I used to live in Acton in West London at the time. It took two or three buses to get to South Mimms from Acton.

'One had to go fairly early in the morning but you just walked in and said "Anyone for Liverpool?" and several hands would go up and you went and joined them in the cab of the lorry and there you are two hundred miles up the road to Liverpool.

'They enjoyed the company on a long journey; they were always alone and they liked to have somebody to talk to. So I knew the route along the old A5 up to Liverpool quite well. The motorway wasn't built then.

'It didn't cost any money – well, it did cost some cigarettes. I smoked at the time anyway and I always had plenty of spare packets of fags to give them. I was in my twenties. In those days you could do it, it was safe.'

9 July

It's one of those surreal moments. It's 9.30 in the evening. The car park is busy. A convoy of white vans waits. Inside it's empty. Apart from a few smokers down their end the main seating area is empty apart from myself and two sad looking souls.

*

We can never be absolutely sure that a service area will never be a destination, at least for a few people. (House of Commons Hansard, April 3, 2001)

*

10 July

The car park is littered with half-emptied plastic fizzy drink bottles. Some contain liquid the colour of urine. A bright red ribbed condom (used) lies alongside assorted pairs of yellow rubber gloves.

Slumbering quietly in cubicle 7 I hear someone whistling 'The Star-Spangled Banner' further along. Should I stand and place my hand on my heart?

Bump into David Seaman, the Arsenal goalkeeper, coming out of the gents' toilet – or is it? His ponytail sways behind him like a drunk.

I join the queue at Burger King. I have to wait ten minutes to place my order for a Spicy Bean burger and fries.

'Do you want the meal, sir, with Coca-Cola?' A spotty young man in his late teens informs me that milkshakes are no longer available. I take a bottled freshly squeezed orange juice instead of the fire bucket size Coca-Cola which comes as part of the meal.

It was a tough call placing the order. The Chicken Whopper Lite with a succulent three ounce of whole chicken looked promising. But then so did the Extra Large Double Whopper with double the beef.

The four counter staff look bored out of their heads but give the impression of some experience in the nuances of critter culinary pleasures.

KFC next door are queuing nineteen deep. Families jostle with singles.

The puddings look tempting. A choice between the Diddy Donuts and a Burger King Cone. I toss a coin, which disappears over the edge of the counter. Fate has intervened so I take neither.

A coachload of pre-pubescent girls and gawky boys fight in the queues.

I retreat to a quiet table with my tray of food.

Manage to drop some tomato sauce down my white shirt.

Wiping it off with a tissue only makes it worse. I look as if a sniper's bullet has hit me.

Both the burger and the fries fill my hunger but leave a film of grease on my tongue which the freshly squeezed orange cannot budge. Feeling desperate I guzzle down a bottle of Coca-Cola from the shop. It does the trick instantly. Knew I should have ordered the set meal!

13 July

Mark is originally from Oldham but now works in Hackney and lives in Turnpike Lane, North London. He has 'passed through' South Mimms a couple of times before. We sit and talk on the outside terrace with its wooden decking and aluminium furniture.

'I'm forty-seven years old and travelled the length of Britain, England certainly, during most of my life as a child and adult. Service stations have changed over the years. I've used many. I can't name them all. But there's one or two significant ones which stay in your mind, mostly those on the M62, as I'm from that neck of the woods, and South Mimms and Watford Gap are kinda like landmarks basically. I tend to remember where they are on the map. But it's also about, I suppose, the experience of the place.

'I've always had this problem with these service stations – like driving down the motorway under pressure to get somewhere at speed, lots of stress, knowing full well at some point of your journey you're gonna have to stop off, have a jimmy riddle, you know you gonna get ripped off to the tune.

'But this one, like, seems modern – you know, relaxing, comfortable, as opposed to other ones which I've gone in just seem to be kind of like a muddle. Lots of people rushing into the toilets, queuing up in the shops, perhaps getting some cash from an ATM, making a phone call, being approached by the AA or RAC, and rushing off back into your car again, filling up with petrol and getting on with your journey.

'So this is quite a pleasant surprise really. Here looks great, attractive, appealing. It's got diverse facilities, and services, and food. It's clean. In fact it's broken the stereotype I had of these places. Like dull, dismal places that I went to reluctantly. But looking at this place it's somewhere you would pop in, and would return to.

'I think it's probably different to what I've experienced before. I've just kinda like had a pleasant journey here, played one or two games, something nice to eat and drink. It's quite a nice day as well here on the patio.

'I'd come back here. But it's all about where's your journey from and to, whether you have to stop off or not. Obviously it's something that's in your mind. You've had that experience, it's been a pleasant one. But I think it's a bit expensive, a lot expensive. But, yea, it's a nice service station as service stations go and there's lots of attractive women to look at.'

15 July

One of the regulars, a social worker (you can tell by the uniform) has had his head shaved. Why? His image has changed. A cross between the evil king in the Flash Gordon remake film and one of the EastEnders' Mitchell brothers. He is also sitting away from his usual seat; he's now in with the suits.

Queuing for my coffee. Danny proudly shows me his engagement ring. He has just returned from Turkey to see his girlfriend and propose. He's getting married next year. Says I will get an invite. I celebrate with a large latté which Danny buys me.

12: A Few References

Augé, M, 'Introduction to an Anthropology of Supermodernity', *Non-Places*, Verso, 1995.

Brittain, F, *South Mimms*, Heffer and Sons, 1931.

De Botton, A, *The Art of Travel*, Hamish Hamilton, 2000.

Hart, F C, *Occasional Papers – No. 4. South Mimms*, Potters Bar and District Historical Society, 1993.

Jeffery, E, *The Bermuda Triangle*, WH Allen, 1975.

Kurtz, I, *The Great American Bus Ride*, Simon and Schuster, 1993.

Lawrence, D, *Always a welcome: The glove compartment history of the motorway service station area*, Between Books, 1999.

Least Heat Moon, W, *Blue Highways. A Journey into America*, Minerva, 1993.

Morris, J, *Destinations*, Oxford University Press, 1980.

Nicholson, J, *A1*, Harper Collins Illustrated, 2000.

Pick, C, *Off The Motorway*, Cadogan Books, 1987.

Sinclair, I, *London Orbital. A Walk around the M25*, Granta Books, 2002.

Printed in the United Kingdom
by Lightning Source UK Ltd.
115861UKS00001B/15

9 781844 013524